KU-330-553

WEATHER RADAR AND FLOOD FORECASTING

Published on behalf of the
British Hydrological Society.

WEATHER RADAR AND FLOOD FORECASTING

Edited by

V. K. COLLINGE
University of Lancaster

and

C. KIRBY
Institute of Hydrology

A Wiley–Interscience Publication

JOHN WILEY & SONS
Chichester · New York · Brisbane · Toronto · Singapore

Copyright © 1987 by John Wiley & Sons Ltd.

All rights reserved.

No part of this book may be reproduced by any means, or transmitted, or translated into a machine language without the written permission of the publisher.

British Library Cataloguing in Publication Data:
Weather, radar and flood forecasting.
 1. Weather forecasting 2. Radar
 meteorology
 I. Collinge, V. R.   II. Kirby, Celia
 551.6′353   QC995

ISBN 0 471 91296 4

Printed and bound in Great Britain

87   02625

Contributing Authors

K. A. Browning   Meteorological Office, Bracknell, Berks, UK

I. D. Cluckie   Dept of Civil Engineering, University of Birmingham, UK

C. G. Collier   Meteorological Office, Bracknell, Berks, UK

V. K. Collinge   Dept of Environmental Science, University of Lancaster, Lancaster LA1 4YQ, UK

F. Dalton   Meteorological Office, Weather Centre (Manchester), Stockport, UK

C. Dobson   Severn-Trent Water Authority, Birmingham, UK

J. R. Douglas   Severn-Trent Water Authority, Birmingham, UK

G. Hill   Meteorological Office, Bracknell, Berks, UK

J. M. Knowles   North West Water Authority, Warrington, Lancs, UK

A. M. Lewis   North West Water Authority, Warrington, Lancs, UK

B. R. May   Meteorological Office, Bracknell, Berks, UK

R. J. Moore   Institute of Hydrology, Wallingford, Oxon, UK

D. H. Newsome   CNS Scientific and Engineering Services, Reading, Berks, UK

G. A. Noonan   North West Water Authority, Warrington, Lancs, UK

M. D. Owens   Severn-Trent Water Authority, Birmingham, UK

D. W. Reed   Institute of Hydrology, Wallingford, Oxon, UK

R. B. Robertson   North West Water Authority, Warrington, Lancs, UK

P. Ryder   Meteorological Office, Bracknell, Berks, UK

G. P. Sargent   Meteorological Office, Bracknell, Berks, UK

G. A. Schultz   Lehrstuhl für Wasserwirtschaft und Umwelttechnik I, Ruhr-Universität, Bochum, Universitätsstrasse 150, 4630 Bochum 1, West Germany

P. D. Walsh   North West Water Authority, Warrington, Lancs, UK

Contents

Preface ix

Part I TECHNICAL DEVELOPMENT OF WEATHER RADAR

1 The Development of Weather Radar in the United Kingdom 3
  V. K. Collinge

2 Cost 72 and Weather Radar in Western Europe 19
  D. H. Newsome

3 The FRONTIERS Project 35
  G. P. Sargent

4 A System for Processing Radar and Gauge Rainfall Data 47
  B. R. May

Part II OPERATIONAL EXPERIENCE OF WEATHER RADARS

5 The Establishment and Operation of an Unmanned Weather Radar
  55
  G. Hill and R. B. Robertson

6 Accuracy of Real-time Radar Measurements 71
  C. G. Collier

7 Weather Radar and a Regional Forecasting Service 97
  F. Dalton

8 An Operational Flood Warning System 109
  G. A. Noonan

viii

Part III   MODELLING RUNOFF USING RADAR DATA

9 UK Flood Forecasting in the 1980s 129
   D. W. Reed

10 Flood Forecasting Hydrology in North West Water 143
   J. M. Knowles

11 Real-time Flood Forecasting in Diverse Drainage Basins 153
   J. R. Douglas and C. Dobson

12 Real-time Rainfall-Runoff Models and Use of Weather Radar
   Information 171
   I. D. Cluckie and M. D. Owens

13 Flood Forecasting Based on Rainfall Radar Measurement and
   Stochastic Rainfall Forecasting in the Federal Republic of
   Germany 191
   G. A. Schultz

Part IV   WEATHER RADAR TECHNOLOGY IN THE FUTURE

14 The Role of Radar and Automated Data Capture in Information
   Systems for Water Management 211
   P. D. Walsh and A. M. Lewis

15 Towards More Effective Use of Radar Data for Flood Fore-
   casting 223
   R. J. Moore

16 Towards the More Effective Use of Radar and Satellite Imagery
   in Weather Forecasting 239
   K. A. Browning

17 Future Development of the UK Weather Radar Network 271
   P. Ryder and C. G. Collier

Author Index 287

Subject Index 289

Preface

This book is based on the proceedings of a symposium 'Weather Radar and Flood Forecasting' held at the University of Lancaster in September 1985. The event marked the completion of the highly successful North West Radar Project, in which the first unmanned weather radar station in the United Kingdom was established and operated to give radar data in real time for use by the Meteorological Office and in the North West Water Authority flood forecasting operations.

Development work in the United Kingdom on weather radar for rainfall measurement and forecasting goes back to the 1950s with research into short term forecasting based on the movement of rainfall echoes. Since that time there has been a continuous programme of research and development, led within public sector by the Meteorological Office and the Royal Signals and Radar Establishment, Malvern with major contributions from the water industry, and by private industry. Major advances have come about as a result of collaborative projects between these parties.

As a result of this substantial and sustained effort the United Kingdom is now a world leader in this field. An operational network of weather radars now covers a large proportion of England and Wales and is soon to be extended further. Satellite data are being integrated in near real-time with radar data for forecasting purposes. The use by water authorities of radar data for flood forecasting is steadily increasing and the considerable potential here is now being recognized. The United Kingdom is also taking the lead in development of a European weather radar network. All of this brings, in addition to the benefits to this country, particularly in flood warning, substantial opportunities in overseas markets for the manufacturers of radar equipment.

The North West Weather Radar Project was promoted, funded and implemented by a Consortium comprising the following organizations:

North West Water
Meteorological Office
Water Research Centre
Department of the Environment
Ministry of Agriculture, Fisheries and Food

This Consortium sought a suitable opportunity to review the results of that project, but it was also seen as an opportunity to bring together and review other related work on flood forecasting and on the use of radar in meteorological forecasting. This brought support particularly from the Institute of Hydrology, which had recently completed a major review of flood forecasting techniques in the UK and also from the British Hydrological Society and the Pennines Hydrological Group. These organizations joined in promoting the Symposium held at the University of Lancaster.

Thus this book, which is based upon the material presented at that Symposium, has been edited and updated into a form which will enable meteorologists and hydrologists to gain not only an up-to-date appreciation of the current technology and capabilities of weather radar, but also an insight into developments currently in progress which are certain to become a part of tomorrow's technology.

Part I
TECHNICAL DEVELOPMENT OF WEATHER RADAR

The introductory two chapters are devoted to the development of weather radar in the UK and its use for flood forecasting purposes, together with the development of a UK network and the work towards an integrated European network carried out as a COST project. Meteorological Office development work is described—on the integration of radar and satellite data for weather forecasting purposes in Chapter 3—and on integrating radar and daily rain gauge data in Chapter 4.

Weather Radar and Flood Forecasting
Edited by V.K. Collinge and C. Kirby
© 1987 John Wiley & Sons Ltd.

CHAPTER 1

The Development of Weather Radar in the United Kingdom

V. K. Collinge

INTRODUCTION

The application of radar technology in meteorology has been the subject of continuous development since radar became available about 40 years ago. Most applications stem from the ability of radar to detect cloud particles, rain drops, snow flakes and ice particles. Apart from its use in precipitation measurement and forecasting, there are other applications; for example, Doppler radar for measuring turbulence, updraft velocities, wind shear and the velocity of sea waves. For a comprehensive presentation of the theory of radar technology and its development world-wide, the reader is referred to Battan, 1973. The research that has taken place in the United Kingdom since the 1940s with associated operational developments has been summarized by Collier, 1984, in a paper designed for an audience of meteorologists.

This chapter reviews the important steps that have occurred in the United Kingdom in developing radar for precipitation measurement and forecasting, and for flood forecasting. It should be remembered that developments have also been continuing in a number of other countries to determine the viability of operational radar systems and networks. Those involved include the USA, the USSR, Finland, Japan, Switzerland, Canada and West Germany.

THE EARLY DAYS

Private industry

Commercial involvement in weather radar development in the United Kingdom goes back to the early 1950s, when Decca Radar Limited produced the

Type 40X band Storm Warning Radar (Bacon, 1985). This equipment was installed at various locations on the international trunk air routes at the time that the first commercial jet aircraft were going into service. The X band had a wavelength of 3 cm, whilst later developments were based on S band (10 cm), and C band (5 cm) systems.[1] The Type 40 had a wide vertical beam of low power, and was able to detect heavier rain showers, and also light rain but only at short range. It was followed in the late 1950s and 1960s by the Types 41, 42 and 43X, which were all X band radars with steadily improving performance. Each of these involved a significant investment of engineering and design effort before going into full production.

The only other early commercial activity to be traced was the system called RAINBOW, developed by Marconi, which was a type of radar demonstrated (it is believed at Farnborough) but not developed further.

In 1965 the ground radar interests (including weather radar) of Decca Radar Limited were sold to the Plessey Company, and a new business, Plessey Radar Limited, was set up. This involved the transfer of development and production facilities and extensive assets to a new site. Plessey Radar took over the production, support and further development of the full range of Decca weather radars. A continuing investment of development resources led to the introduction of a completely new S band radar in 1966—the Type 43S. It was this equipment which was used initially in the Dee Weather Radar Project.

The Meteorological Office

The potential of radar for meteorological purposes was recognized from the early days of radar development (Collier, 1984) and a Radar Research Station was established during the Second World War at East Hill near Dunstable. Early work on cloud physics gave way to examination of the accuracy of short-term rainfall forecasts based on the movement of associated radar echoes. In June 1955 a Decca 3 cm radar was installed on the Air Ministry roof in London as an aid to operational forecasting. Doubts were expressed in 1957, in the Meteorological Office Annual Report of that year, about the value of radar in relation to the expense involved, and in the next 10 years only three more 3 cm radars were installed in the UK for operational forecasting.

Despite these doubts about the technology current at that time, research activities began to develop and expand. The loan of a 3 cm Doppler radar by the Royal Radar Establishment led to the closure of the East Hill site and its transfer to Malvern for absorption into a joint Meteorological Office/ Royal Radar Establishment group. The development of radar theory, the

---

[1] In general the shorter the wavelength the smaller the size of the particles it is able to detect.

use of radar in the study of thunderstorm dynamics and the use of Doppler radar formed the main themes of research up to the mid-1960s.

DEE WEATHER RADAR PROJECT

In retrospect it is clear that this project proved to be the cornerstone in the progression in the UK from rainfall detection to rainfall measurement by radar. Its foundation lay in a research programme initiated in 1966 by the Water Resources Board aimed at developing new methods for operating river regulation systems. At that time the river Dee in North Wales was the most heavily regulated river in the country, with a lake (Bala) and reservoir (Celyn) used to sustain low river flows and reduce flood peaks. Interest within the Water Resources Board in the potential of weather radar as a means towards improving the efficiency of regulation led to a meeting with the Meteorological Office and subsequently with Plessey Radar Limited. In 1967 these three parties agreed to support a project aimed at developing weather radar for the quantitative measurement of rainfall. Plessey Radar Limited supplied the radar on free loan, the Meteorological Office agreed to operate it and meet the running costs, and the Water Resources Board met the capital costs of the project. The project was in the Dee basin, already being developed for the hydrological work in conjunction with the Dee and Clwyd River Authority, and who readily agreed to collaborate in the extra work involved.

The objectives of the project were:

1. to investigate the accuracy with which areal precipitation can be measured in a hilly (mountainous) area; and
2. to develop a real-time system for the measurement of areal precipitation on time and space scales appropriate to the hydrological requirements for water management and river regulation.

In addition it was hoped that an examination could be made of ways in which radar data might be used to improve the short-period quantitative forecasting of precipitation.

Trials with a mobile radar led to the choice of a site on the Llandegla Moors near Wrexham. The radar equipment was largely installed by March 1971 and comprised a standard Plessey Type 43S (10 cm) set with only minor modifications. The beam width was 2°. In addition there was an automatic echo signal analyser to digitize the video output into 32 levels and average them into 'bins' which were 300 m in range and 2° in azimuth. A magnetic tape recorder was used for off-line data analysis by the Meteorological Office at Bracknell. In 1973 the radar was converted from S band to C band, which also reduced the beam width. This was expected to improve performance by reducing permanent echo and reducing the duration of interference by bright band echo, and these expectations were realized. A network of 60 tipping bucket raingauges with magnetic tape recorders was used for comparison with the radar output; maintenance and operation of these gauges required substantial effort. The gauges were set with the rim at ground level in a pit, and surrounded by a non-splash lattice.

As a result of the reorganization under the Water Act of 1973 the Water Resources Board was abolished and its responsibilities for the project were taken on by the Central Water Planning Unit, the Water Research Centre and the Water Data Unit. The Dee and Clwyd River Authority became part of the Welsh National Water Development Authority which agreed to continue support. The radar work continued until October 1976, merged with other work on hydrological forecasting in the re-named project 'The Dee Weather Radar and Real-Time Hydrological Forecasting Project'. Data processing, transmission and display facilities developed as part of the project at Malvern on radar network development (see below) were tested in parallel with the existing Dee Project data processing system towards the end of that project in 1975.

The Dee Projects were discussed at a Conference in Chester in December 1975 entitled 'Weather Radar and Water Management', and were fully documented in the final project report (Central Water Planning Unit, 1977). The results of the radar studies were also published, as for example, in Harrold *et al.*, 1974 and Collier *et al.*, 1975. The main conclusions derived from the projects are summarized below as an extract from the final report:

1. The measurement of areal precipitation by radar has been established as effective in hilly country as well as over lowland areas. The method provides data over an extensive area in real time and includes information on intensity distribution not readily available from any other source. Except for relatively small areas it is cheaper than a comparable tele-metering raingauge network.
2. Radar networks have great potential for exploiting the advantages of the radar measurement for the many activities interested in current and immediate future precipitation. These activities include meteorological and hydrological forecasting, electric power generation, urban sewerage management, civil aviation, agriculture and the construction industries.
3. Precipitation forecasting, in both time and amount, can be of very great value to hydrological forecasting.
4. On-line hydrological forecasting has been shown to be feasible.
5. In the general context, a major contribution to the success of the project was the system of continuous analysis of the data as soon as possible after acquisition, so enabling a close watch to be kept not only on the operational system, but on the effective achievement of the objectives.

6. In the wider context, and with the possibility of short-period quantitative rainfall forecasts, improved control rules for reservoirs are likely to be achieved to realise worthwhile benefits.

DEVELOPMENTS IN THE MID-1970s

Opinions had hardened on the choice of frequency (Taylor and Browning, 1974), in favour of the C band (5 cm wavelength, 5.6 GHz frequency) as being the optimum for use in temperate latitudes. The cost of an S band aerial was thought to be prohibitive, whilst attenuation due to rain would prevent the use of an X band radar. Attenuation at C band, although significant in extensive heavy rain, can be corrected for in the on-site computer.

Plessey Radar's experience of equipment operation in the Dee Weather Radar Project and elsewhere, and continuing discussions with customers, showed the need for a new range of C and S band radars to meet the developing needs of hydrological and meteorological services. An extensive development programme was undertaken in the period 1974–76 leading to the development and production (1976 onwards) of the modular family of Plessey weather radars, the types 46C, 45C and 45S. The cost of this development programme was £300 000–£400 000 at 1975 prices (Bacon, 1985). The most important advance was in the achievement of an un-manned system, of critical importance in reducing operating costs. This required extensive software development which was carried out by the Meteorological Office, in addition to Plessey's work on the hardware. Without these investments, which produced a much more advanced system than the Type 43 and particularly an unmanned system, it is highly unlikely that the North-West Radar Project would have proceeded, and certainly it would have been delayed by several years.

TOWARDS A NATIONAL NETWORK: 1971–81

At about the time that the Dee radar was installed, Bulman and Browning (1971) prepared a preliminary report at the request of the Director-General of the Meteorological Office on the possibilities for a national weather radar network, and this became the first major step towards the goal which is now fast being realized. The authors envisaged an eventual network of twelve C-band radars to cover the whole of the British Isles. Un-manned radars were by then in use for military purposes, and so were expected to become available for civilian purposes including weather forecasting. At this time the possible benefits of such a network received little consideration, the emphasis being on solving the problems of data transmission, processing and display.

This report led to a collaborative project between the Royal Signals and Radar Establishment (RSRE) and the Meteorological Office at Malvern to develop data-processing facilities for a network of weather radars with a centrally composited television-type display. In view of the size and cost of the task a step-by-step approach was adopted, starting with data from a single radar site. By late 1973 it had been demonstrated that data could be processed in real time, transmitted from the site and displayed remotely. Further work on linking several radars followed (Taylor and Browning, 1974; Taylor, 1975), successfully demonstrating the feasibility of operating a network of three research radars—Castlemartin (South Wales), Defford (Worcestershire) and Llandegla (Dee catchment, North Wales).

Further impetus from the Dee Weather Project came through what became known as the Operations Systems Group (Water Resources Board, 1973). This group studied in some detail operational requirements and systems needed, and their costs, but avoided the difficult matter of benefits. The report recommended that consideration be given to establishing a radar network to cover the United Kingdom, and to be co-ordinated where possible with any similar networks in neighbouring countries. The benefit aspect only came under serious scrutiny as an outcome of the conference in Chester in 1975, which discussed the results of the Dee Weather Radar Project. An assessment of the potential benefits of a national weather radar network was undertaken by the Central Water Planning Unit and the Water Research Centre (Bussell *et al.*, 1978). Potential benefits were estimated at £22 million p.a., mainly from:

> Construction industry (£10 million p.a.)
> Agriculture (£6.5 million p.a.)
> Water industry (flood warning) (£3.76 million p.a.) and
> Transport (£1.8 million p.a.).

The successful linking of several radars referred to above, coupled with other work, and with the availability of $\frac{1}{2}$ hourly cloud imagery from the weather satellite Meteosat, led to a further Meteorological Office project, the Short-Period Weather Forecasting Pilot Project (Browning, 1977) which began in 1978. This 5–8-year project, based at the Meteorological Office Radar Research Laboratory at Malvern, was intended to develop methods to give improved forecasts of precipitation and wind for the period 0–6 hours ahead, particularly through the use of weather radar and satellite data.

Work up to 1981 was largely aimed at establishing facilities to provide mesoscale observational fields of cloud and precipitation, initially over part of the United Kingdom. A key element was the establishment of a network of four radars providing composited data in real time to Malvern. These radars were as shown in Table 1.1.

The radars at Cambourne and Upavon were of old design and give data

Table 1.1

Site	Date	Notes
Cambourne, Cornwall	1978	Old, manned S band (10 cm wavelength, 2° beamwidth). Sites manned by technicians.
Upavon, Wiltshire	1979	
Clee Hill, Shropshire	1979	C band (5.6 cm wavelength, 1° beamwidth). Unmanned but Civil Aviation Authority staff on site. Radar relocated from Dee.
Hameldon Hill, Lancashire	1979	North-west Radar Project.

of less accuracy than could be obtained by modern equipment, as well as being less reliable. Also the sites, selected partly for logistic support reasons, were not the optimum choice. The site at Clee Hill was selected for the radar equipment from the Dee Project because of the good coverage of a large area of Wales and the Midlands. The fourth radar was installed at Hameldon Hill, near Burnley in Lancashire, in 1979. It was the first of the new generation of unmanned C band radars, installed for the North-West Radar Project (see below). These four radars comprised the first basic network in the United Kingdom (Figure 1.1) on which later developments have been based.

THE NORTH-WEST RADAR PROJECT

By 1975 the Dee Weather Radar Project was clearly indicating the conclusions given above. However, the radar was manned continuously and so had very high running costs, the data processing was largely off-line and the problems of calibration had been only partly resolved. During the Chester conference in December 1975, held to discuss the project findings, discussion between the author and Dr Marshall of North West Water Authority (NWWA) revealed an exciting possibility for further development work. The NWWA were at that time planning a major telecommunications network which was to include numerous telemetering rain gauges for flood warning purposes. The opportunity therefore existed for a reduced network of these rain gauges coupled with an unmanned weather radar, as a further development project aimed at achieving an operational flood forecasting system. It is to the credit of NWWA that they responded to this possibility in a positive and forward-thinking way.

Thus in 1977 the North-West Radar Project (NWRP) was initiated, and

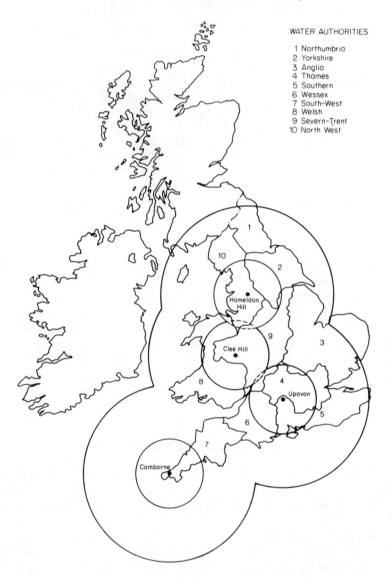

WATER AUTHORITIES

1 Northumbria
2 Yorkshire
3 Anglia
4 Thames
5 Southern
6 Wessex
7 South-West
8 Welsh
9 Severn-Trent
10 North West

Figure 1.1.   Weather radars 1980. Circles at 75 km and 200 km radius

was funded by a consortium comprising the Meteorological Office; North-West Water Authority; Water Research Centre; Central Water Planning Unit and the Ministry of Agriculture, Fisheries and Food. It is not appropriate to describe the project here since full details are given in Chapters 5, 6, 8 and 10 and in the final project report (North-West Water Authority, 1985).

However, it is useful to summarize the key developments achieved by the project:

1. The establishment of the first unmanned radar in the UK (Hameldon Hill, Lancashire) producing precipitation data in real time.
2. Performance evaluation of that radar, proving that the concept of an unmanned radar was sound and that operational reliability was very high.
3. Integration of the radar into the NWWA Regional Communications System.
4. Integration of the radar data with that from other weather radars as a prototype UK network.
5. Further development of quantitative precipitation forecasting.
6. Development of hydrological forecasting methods.
7. Operational use of radar data in flood forecasting in north-west England.

TOWARDS A NATIONAL NETWORK: 1981 ONWARDS

The progress achieved by 1981 with the North-West Radar Project, and the successful networking of four radars, led to the setting up in December 1981 of a joint Meteorological Office/National Water Council Working Group to consider the implication for water authorities of a national radar network. The group comprised representatives from water authorities, the Meteorological Office and the Ministry of Agriculture, Fisheries and Food. Its terms of reference were: 'To examine the broad outlines of a National Weather Radar Network that would fit Water Authority requirements in England and Wales, and to consider the implications for water authorities as far as manpower and costs were concerned.' In its report (National Water Council, 1983) the group confirmed that operational experience, particularly of an unmanned radar, had demonstrated significant advantages to water authorities in flood prediction and warning, and endorsed the development of a national weather radar network.

It envisaged a network of eleven or twelve radars which would provide quantitative coverage (radius 75 km) over most of England and Wales. For the purposes of economic evaluation it was assumed that the full capital cost of eleven radars would have to be met (although in practice several already existed) with a life of 10 years. At a 5 per cent discount rate this gave a total net present value (NPV) of £8.31 million, with the water industry assumed to be providing 50 per cent of this, i.e. £4.15 million.

The study of potential benefits was substantial in so far as flood damage reduction was concerned, and gave a net present value (again with 5 per cent discount rate) of £8.98 million. This was without the introduction of FRONTIERS, a project aimed at improving precipitation forecasts through the combination of radar and satellite data, and described in Chapter 3. On these assumptions the ratio of

$$NPV \text{ of benefits/NPV of costs} = 2.15$$

With the full introduction of FRONTIERS this would rise to 3.35.

It is also interesting to note that no benefits were ascribed to reductions in traffic dislocation from flooding, which accounted for 40 per cent of the benefits estimated in the earlier study (Bussell *et al.*, 1978). Because this 1983 study was aimed at the water industry, no evaluation of other benefits was attempted beyond a qualitative description.

By 1984 a further major development had taken place. Through a consortium comprising the Meteorological Office, Greater London Council, Thames Water Authority and Southern Water Authority a fifth unmanned C band radar was being commissioned at Chenies, Buckinghamshire, to cover London and South-East England (Figure 1.2). Data from an S band radar at Shannon Airport in the Republic of Ireland were also being incorporated into the UK composite pictures.

These five radars in England and Wales operate continuously and provide data directly to users on both 2 km and 5 km grids and over predefined river subcatchments every 15 minutes. Data are also transmitted from each site to a network centre at the Meteorological Office, Bracknell. The data are composited to form a $128 \times 128 \times 5$ km or $256 \times 256 \times 5$ km image of inferred precipitation intensity every 15 minutes, which is distributed to water authorities and Meteorological Office users.

Each radar site has its own dedicated mini-computer system. The software for these computers and for the network centre computer has been developed by the Meteorological Office from earlier software developed by the MO/RSRE team at Malvern. A variety of different processing tests are carried out at the radar site. At the network centre complex communication and compositing software has been needed to process data simultaneously from up to sixteen radar sites, and from various other sources. All this software development has taken about 30 man-years of effort. These data processing systems are described in detail in Chapter 17.

The future

A number of further developments are in hand and are also described in Chapter 17. The actual and forecast future growth in radar coverage is illustrated in Figure 1.3. The growth curve on an area basis assumes coverage up to 75 km radius. In benefit terms, the lower curve assumes no benefits beyond 75 km radius, and the upper curve assumes that 50 per cent of the possible benefits beyond the distance are achieved.

Figure 1.2. Weather radars 1984. Circles at 75 km and 200 km radius

Meanwhile interest has extended to Scotland, with an evaluation started in January 1984 and completed mid-1985, carried out by the Scottish Development Department, the Meteorological Office and other agencies (police, drainage, highways, electricity). This has demonstrated a case for three weather radars in Scotland, though in marked contrast to England the

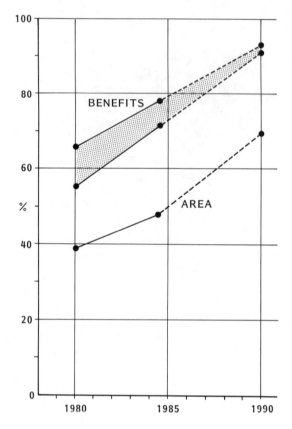

Figure 1.3.   Growth in quantitative weather radar
coverage, England and Wales

benefits are expected to derive mainly from more economic treatment of
roads with salt in winter (£1.2–3.7 million p.a.) followed by more efficient
pesticide application (£0.6 million p.a.) and benefits to the construction
industry (£0.4 million p.a.), with flood forecasting showing negligible benefits
at £0.1 million p.a.

WATER AUTHORITY USE OF WEATHER RADAR DATA

As a consequence of the various development projects and the gradual
development of a prototype network, there has been a steady growth in the
use made by water authorities of radar data, which is summarized below.
This is in addition to the substantial direct involvement by Welsh Water and

North-West Water in the two development projects described above, and more recently by Thames Water and Southern Water in the operational radar at Chenies, Buckinghamshire, which covers the London area.

Yorkshire Water

This authority first received data from the Hameldon Hill radar station, by landline via the NWWA Divisional Office in Warrington, in December 1982. A Jasmin mark III A processor was used to assemble the data and store nine quarter-hourly pictures. The data are received and displayed in the Rivers Division Head Office in Leeds, where they are used qualitatively at flood times to provide an areal interpretation of quantitative data received from telemetry raingauges. Verbal descriptions of the weather radar display, including the extent, intensity, and movement of storms, are given by telephone to flood operations offices. The radar data are of particular use when the rivers are running bank-full, when knowledge of the likelihood of significant additional precipitation can lead to improved forecasting of floods and can affect decisions on the staffing of flood operations offices and the deployment of labour to areas at risk.

The traditional magnetic tape recording system was considered to have drawbacks and an alternative system incorporating a PAL encoder, a timer, a domestic video recorder and a second display monitor was installed for recording and playing back displays. The system has been successful but is far less versatile, and produces a recording with poorer definition than the microcomputer-based systems now available.

The Meteorological Office has written additional software for insertion in the Hameldon Hill computer to give precipitation totals for 57 Yorkshire Water subcatchments. The resulting data will be accessed from the datastream using a purpose-made printer. As part of a major hydrological data handling project it is proposed to establish a regional flow forecasting centre for which real-time catchment models will be developed. The radar data will be a major input, together with hydrological data retrieved by a new telemetry system. The data will be received and assembled by a computer which will interface with the catchment modelling computer.

Welsh Water

Following involvement in the latter stages of the Dee project, proposals in 1970 for the authority to receive data from the Clee Hill radar were considered but not pursued. It does consult Severn-Trent Water Authority and Cardiff Weather Centre (which has a radar) when important forecasting situations arise. The authority has joined the consortium which in 1985 was promoting two new radars (Exmoor and Pembroke).

Severn–Trent

This authority's interest started with the Chester Conference in December 1975, following which in 1976 it purchased a terminal for receiving data from the Dee radar (Llandegla). It was used in the Malvern office for operation control work in the Severn basin—regulation and flood control. This gave good experience and proved a valuable tool in controlling reservoirs in drought conditions, since 1976 produced the 'driest five consecutive months for 250 years.

Subsequently the Dee radar was refurbished and moved to Clee Hill in 1979, which the authority encouraged since it gave good cover of about two-thirds of their area. Data from Clee Hill were then used and handled via a Jasmin microprocessor-based printer at their Malvern office giving rainfall totals for over 100 subcatchments.

Subsequently the authority has carried out a very detailed programme to assess the value of weather radar to it, with the results brought together in a substantial and detailed report (Severn–Trent Water Authority, 1986). It would appear that the authority can expect to become further involved in the development of a national weather radar network and is likely to participate in the proposed radar in East Anglia (Lincoln). Meanwhile in 1985 it purchased and installed in their headquarters a new terminal manufactured by Software Sciences. The inclusion of radar data in real-time flow forecasting models awaits regionally available fully calibrated quality controlled data from the national network. Plans are progressing to disseminate data to divisional users.

Wessex Water

Wessex Water Authority made use of the single site Upavon Weather Radar data at its Bristol Avon Division's office at Bath, duplicated at the Regional Operations Centre at Bristol shortly after it became available. These data were received as both a degraded picture for visual display and as sub-catchment totals for post event analysis. With the provision of the four site composite data (Cambourne, Upavon, Clee Hill and Hameldon Hill) the authority then received these data for its Somerset Division at Bridgwater on the boundary of three of the four radars. This service, duplicated at the Regional Operations Centre at Bristol, was provided concurrent with the same service to the Bristol Weather Centre. All sites used the Jasmin unit with cassette tape archive.

In February 1985 the authority replaced the Jasmin unit at its Regional Operations Centre with a Logica Vitesse system to facilitate rapid replay and zoom of composite images. All locations use the radar qualitatively for real-time flood warning; only the Bristol Avon Division uses quantitative data, as subcatchment totals for post-event analysis.

During 1985/86 the authority evaluated alternative real-time flood pre-diction models that use the radar data as the direct precipitation input. The results of this research will determine the future direction of modelling and the use of radar and telemetered data for flood warning purposes to be implemented when the authority enjoys full quantitative coverage from the proposed new radars in Pembroke and on Exmoor (Wessex Water is a member of the Consortium promoting these new installations).

South-West Water

This authority purchased a Jasmin unit in 1978, linked to the Cambourne radar, for use in their flood warning office. The coverage is generally the whole of Cornwall (except the Tamar catchment) where quantitative data can be relied on and a network of telemetry gauges is used for the remainder of their area. Experience has established the colour display from Cambourne as a most important forecasting aid, giving a quick indication of the approach and onset of rainfall and of its pattern, though the data are not used directly in hydrological models. In 1981 the Jasmin unit was replaced by a system based on a BBC microcomputer to give improved storage, display and replay facilities. Meanwhile the authority is actively involved in promoting two new radars (Exmoor/Pembroke). The intention would be to then take the composite picture from Bracknell rather than the single-source Cambourne data.

CONCLUSIONS

Impressive advances have been made over the past 20 years in weather radar technology, and the achievement of an integrated network covering England and Wales is now within sight. It has put the UK in the forefront of the technology at an international level, is contributing to important advances in short-term precipitation forecasting, and has opened up an important export market. The new range of weather radars marketed by the Plessey Company has been supplied to a number of overseas countries including Saudi Arabia, Cameroun, Italy, France and Brazil, and the company (or its predecessor) has supplied world-wide over 190 weather radars.

All of this could not have been achieved without the closest collaboration between Government agencies, private industry and the water industry. It also required in the early days a commitment to a large research expenditure without any meaningful estimate of the potential benefits being possible. The way is now clear in this country for the water industry to take full advantage of the data that are and will become available as a result of this network and other data, including satellites.

ACKNOWLEDGEMENTS

The author is grateful for the assistance received through the provision of information, reports and papers from P. Bacon (Plessey Radar), C. Collier (Meteorological Office), North-West Water, J. Tinkler (Wessex Water), J. Neat (Yorkshire Water), G. Bull (South-West Water), W. L. Jack (Welsh Water), R. Goodhew (Severn–Trent Water).

REFERENCES

Bacon, P. J. (1985). Private communication.
Battan, L. J. (1973). *Radar Observations of the Atmosphere*. University of Chicago Press, Chicago and London.
Browning, K. A. (1977). The short-period weather forecasting pilot project. *Met. O. RRL Research Report No. 1.*
Bulman, P. J. and Browning, K. A. (1971). National weather radar network. *RRE Report*, July, 17 pp.
Bussell, R. B., Cole, J. A. and Collier, C. (1978). The potential benefit from a national network of precipitation radars and short period forecasting. *Central Water Planning Unit*. 39 pp + 1 appendix.
Central Water Planning Unit. (1977). *Dee weather radar and real time hydrological forecasting project*. Report by Steering Committee, Central Water Planning Unit, London, HMSO.
Collier, C. G. (1984). Radar meteorology in the United Kingdom. *Proc. 22nd. Conf. on Radar Met., Zurich, Switzerland. AMS Boston*, 10–13 Sept., pp. 1–8.
Collier, C. G., Harrold, T. W. and Nicholass, C. A. (1975). A comparison of areal rainfall as measured by a raingauge calibrated radar system and raingauge networks of various densities. *Proc. 16th Conf. on Radar Met., Houston, Texas, AMS Boston*, 22–24 April, pp. 467–72.
Harrold, T. W., English, E. J. and Nicholass, C. A. (1974). The accuracy of radar-derived rainfall measurements in hilly terrain. *Quart. J. R. Met. Soc.* **100**, 331–50.
National Water Council (1983). Report of the Working Group on National Weather Radar Coverage, National Water Council/Meteorological Office, 30 pp. plus appendices.
North-West Water Authority (1985). North-West Weather Radar Project; Report of Steering Group.
Severn–Trent Water Authority. (1986). *An evaluation of radar for the measurement and forecasting of rainfall*. Severn–Trent Water Authority, Birmingham.
Taylor, B. C. (1975a). A system for real-time processing, transmission and display of radar-derived radar data. *Proc. Water Research Centre/Royal Radar Establishment Conference Weather Radar and Water Management, Chester*, December 1975.
Taylor, B. C. (1975b). A mini-network of weather radars. *Proc. 16th Conf. on Radar Met., Houston, Texas, AMS Boston*, 22–24 April, pp. 361–3.
Taylor, B. A. and Browning, K. A. (1974). Towards an automated weather radar network. *Weather*, **29**, 202–16.
Water Resources Board (1973). *Dee Weather Radar Project*. Report by the Operations Systems Group on the use of a radar network for the measurement and quantitative forecasting of precipitation. 21 pp + 9 appendices.

Weather Radar and Flood Forecasting
Edited by V.K. Collinge and C. Kirby
© 1987 John Wiley & Sons Ltd.

CHAPTER 2

Cost 72 and Weather Radar in Western Europe

D. H. NEWSOME

INTRODUCTION

Traditionally, data on precipitation have been gathered from networks of raingauges which are situated in many different locations. Some of these are readily accessible; others, perhaps on tops of mountains, may be less so. Some are interrogable, some are read daily, others weekly or monthly. The cost of maintaining a network of raingauges is not inconsiderable. Using weather radar, however, it is now possible to detect and measure all forms of precipitation to an acceptable accuracy for operational management purposes, in near real time, *from a single location*. Moreover, in association with a computer, data from weather radars can be processed to provide areal totals of precipitation over subcatchments or entire catchments of river basins. This information enables streamflows and levels to be predicted, which can be particularly useful in potential flooding situations.

A technique has been developed in the United Kingdom to integrate data from a network of radars on an operational basis (i.e. in real time) enabling information to be made available about intensity of precipitation over large areas. Similar development work has also been carried out elsewhere, e.g. Switzerland. Over the past few years the UK Meteorological Office has also designed and implemented a software system which will merge precipitation data obtained from the network of weather radars with satellite data on cloud cover and cloud-top temperatures. Forecasters using this information can prepare weather forecasts which are more accurate than those compiled at present, particularly with regard to the timing of the onset or cessation of precipitation. Additionally, the detection of hazardous meteorological phenomena is significantly improved.

During the past 5 years representatives of thirteen countries have been planning an integrated network of weather radars covering Western Europe and devising protocols, standards and methods whereby data can be usefully exchanged. In 1985, the last year of the present COST[1] 72 project, a pilot scheme has been in operation merging data from weather radars in the Republic of Ireland, the United Kingdom, France and Switzerland. This chapter outlines the history of the COST 72 project 'Measurement of precipitation by radar', describes its pilot project and indicates the benefits that may be expected from an integrated network of weather radars in Western Europe.

COST 72

COST already has some notable successes to its credit, with the establishment of the European Centre for Medium Range Weather Forecasting, Shinfield Park, Reading, being a particularly good example. COST 72 is the project dealing with the measurement of precipitation by radar and, *inter alia*, is concerned to establish the utility of an integrated network of weather radars for Western Europe for meteorological and hydrological forecasting, to recommend the data to be exchanged between countries and to suggest the pattern an integrated network might take. A Memorandum of Understanding for the implementation of a European research project in this field was signed by ten of the nineteen countries which participate in COST projects. They are: Denmark, Finland, France, Federal Republic of Germany, Italy, The Netherlands, Portugal, Sweden, Switzerland and the United Kingdom. Other countries expressing interest and attending COST 72 meetings from time to time are: Austria, Belgium, Greece, Ireland, Spain and Yugoslavia. (The remaining countries of the COST group are Luxembourg, Norway and Turkey.)

This Memorandum defined the objectives as:

1. To study the technical and financial aspects of a co-ordinated approach to a European weather radar network.
2. To improve the quality of radar data and the correlation between radar data and meteorological phenomena, primarily in the form of precipitation levels and totals, for short-period forecasting and other purposes, for potential users such as meteorologists, hydrologists, aviation, the con-

---

[1] COST is an acronym for 'European Co-operation in the field of Scientific and Technical Research'.

struction industry and agriculture, taking into account especially the giving of warnings aimed at saving life and property.

3. To optimize the cost–benefit ratios of national networks, of lateral networks and, ultimately, European networks.
4. To study the possibility of standardization of radar-observing systems with the aim of facilitating the economic production of equipment on a European basis.
5. To examine the possibilities of combining weather radar data with cloud data from meteorological satellites.

Work designed to meet these objectives is now almost complete, and the final report has been prepared. A conceptual system for the dissemination of radar/satellite derived information has been defined (Figure 2.1) together with user requirements in so-called normal and abnormal weather conditions[2] (Newsome, 1981). The user requirements in abnormal weather conditions are set out in Table 2.1; those for normal weather conditions follow a similar pattern but are less exacting. The data to be exchanged between neighbouring countries (high resolution in space and time) and between countries generally (low resolution in space and time), has been considered and existing and planned work in COST countries on the merging of data from two or more radars and on the integration of data from radar networks with those from other meteorological sources, notably satellites, has been reviewed by working groups of the COST 72 Co-ordinating Committee. This work culminated in the mounting of a pilot project (Collier and Fair, 1985) to exchange data between a number of countries to arrive at a better estimate of the utility of so doing between neighbouring countries and between countries generally.

A number of Operational Assessment Centres (OACs) staffed by meteorologists were to be established for this purpose. In the UK the designated OAC is the Meteorological Office at London's Heathrow Airport. Other centres in meteorological offices in Austria, Finland, France, The Netherlands, Sweden and Switzerland have also been established. This spread of OACs has enabled as broad a spectrum of opinion to be canvassed as is practically possible within the limitations of the pilot project.

THE COST 72 PILOT PROJECT

For a number of reasons, frequently lack of staff, equipment or financial resources, it took 3 years from the inception and planning of the pilot project

---

[2] For the purposes of this paper these are defined to be: *normal conditions*—intensities and duration of precipitation which do not cause the user significant problems; *abnormal conditions*—intensities and/or durations of precipitation which cause the user significant problems and may cause him to take precautions to minimize the adverse effects of such events.

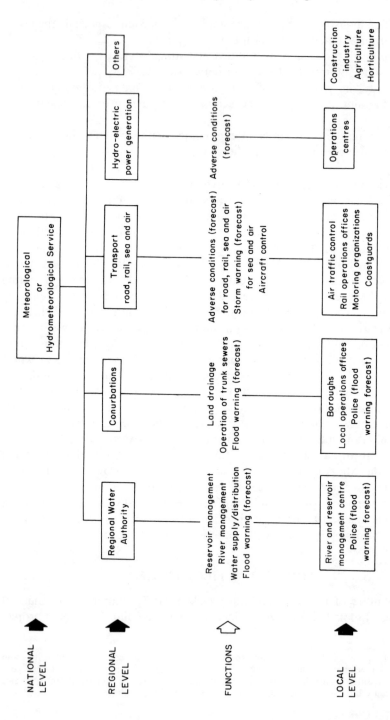

Figure 2.1.   Conceptual information diagram

Table 2.1. User requirements ('abnormal' weather conditions)

	Meteorological		Hydrological			Agriculture		Transport	Construction
	Central (principal) office	Subsidiary (local) office	River basin operation centre	Urban sewer operation	Hydro-electric power operation	General farming	Horticulture	Road/rail transport authorities	Major project offices
1. Area of interest	National	Regional	River basin	Urban area	River basin	Regional	Local	Regional	Local
2. Areal resolution (km²)	50	2	50	2	50	50	2	50	2
3. Precipitation reporting	1 h	5 min	15 min	5 min	30 min	1 h	1 h	1 h	1 h
3.1 Frequency of reporting 3.2 Period over which precipitation is totalled 3.3 Resolution of reported precipitation	Integrated totals over period between reports								
(a) Number of levels of intensity (mm/h)	6 (selected from 127)	127	25	25	25	8	8	8	8
(b) Scale (mm/totalled)			0–25.0 mm increments of 1 mm		0–25.0 mm, 1.0 mm increments	0–25.0 mm, increments of 3.0 mm			
4. Precipitation forecasting									
4.1 Frequency of forecast message	1 h	15 min	15 min	15 min	30 min	3 h	3 h	1 h	1 h
4.2 Forecast period ahead	36 h	3 h	3 h	1 h	24 h	3 h	3 h	3 h	3 h
4.3 Time resolution of forecast message	1st 6 h 1 h Remainder 6 h	5 min	15 min	15 min	30 min	1 h	1 h	1 h	1 h

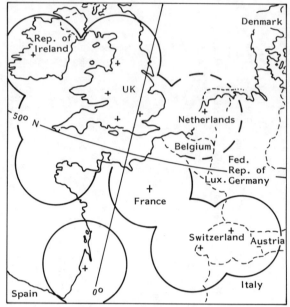

Figure 2.2.   The pilot project image area (polar stereographic projection)

to the operational stage which commenced in Spring 1985. It was confined
to the exchange of low-resolution radar data and radar/satellite data from
one country to several others. The United Kingdom was designated as the
principal data source, which involved receiving data streams from three
countries—Switzerland, France and Ireland—which were designated as secondary data sources. The data streams are received in a number of different
formats, processed into a standard format, merged and combined with
Meteosat data to form the pilot project image, or COST image. The composite picture is then transmitted either directly by the PSTN (public switched
telephone network), using an autodial facility, or via the WMO global
telecommunications system (GTS) to the other participating countries. The
extent of the pilot project area is shown in Figure 2.2 and the central
computer configuration used is shown in Figure 2.3.

Production of the pilot project image (or COST image)

After site processing (James, 1981), data from UK and Irish radars are
produced as images on a 5 km grid aligned to the UK National Grid
Projection (Transverse Mercator) without difficulty. However, the French
and Swiss data are on approximately 2 km × 2 km grids, both use different

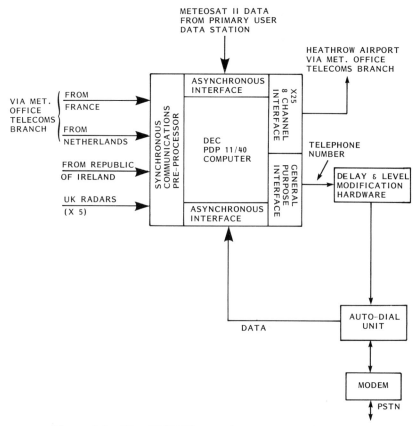

Figure 2.3. The COST 72 central computer configuration

intensity level slicing schemes from the UK, and each has its unique trans-mission code. In addition, the satellite data are acquired as a space-view image covering the Western European countries and Northern Atlantic areas. Because of the angle of view from Meteosat II, equal areas of the earth's surface appear quite differently on the image, depending principally on the latitude, but also on the longitude of the area in question. To overcome these difficulties it was decided early on in the pilot project to convert the image to a polar-stereographic map projection.

After decoding the data to a common format, an initialization program is run to set up all the variables appropriate to the particular projection used. Two files are prepared, one containing radar data and the other satellite data, and these are used to cross-reference the x–y co-ordinates of each pixel in the pilot project image to the x–y co-ordinates in the appropriate radar or satellite image. Associated with the cross-reference of the radar image is

Weather radar and flood forecasting

Table 2.2. Level slicing used initially for pilot project images (Collier and Fair, 1985)

0	0.0 to 0.125	>−3 to −00
1	>0.125 to 1	>−3 to −27
2	>1 to 4	>−27 to −35
3	>4 to 8	>−35 to −42
4	>8 to 16	>−42 to −50
5	>16 to 32	>−50 to −55
6	>32 to 100	>−55 to −60
7	>100	>−00 to >−60

Table 2.3. Level slicing now in use for the pilot project images (Collier and Fair, 1985)

Slice value	Radar (mm n^{-1})	Satellite (°C)
0		>−15 to −00
1		>−15 to −45
2		<−45 to −00
3	>0.3 to 1	
4	>1 to 3	
5	>3 to 10	
6	>10 to 30	
7	>30 to 00	

an indication as to which radar image is to be used at each point in the pilot project image. The output of the 11/40, is then used as a look-up table in all subsequent operational runs.

Each time new data become available, usually every hour, they are re-projected to polar-stereographic co-ordinates using the output files created by the initialization program run. All radar data are converted to an approximate 5 km × 5 km grid when necessary by averaging an appropriate number of values, and are then level-sliced according to the scheme set out in Table 2.2.

Satellite infra-red data are level sliced according to the scheme proposed by Negri and Adler (1981) also shown in Table 2.2. While this scheme gives acceptable results in the convective situations for which it was intended, it is less satisfactory with frontal rainfall. Where no radar data are available, the value from the appropriate *x–y* pixel in the satellite image is level sliced and stored in the pilot project image. More recently, based on experience of these images in Switzerland, Dr J. Joss of the Observatorio Ticinese, Locarno-Monti, suggested that the data should be level sliced so that different groups of levels are allocated to the radar and satellite data, as shown in Table 2.3. This enables satellite data to be used wherever there is no precipitation indicated by radar, and alleviates the large discrepancies of well over an order of magnitude which often occur between satellite and radar data. The data are checked to ensure that all data times are within 30 min of each other. Any data more than 30 min old are rejected. Finally, a scan is made of the image output, and wherever a change from radar to satellite data (or vice-versa) along a line is detected, a marker is inserted in the image. The images are then disseminated to Switzerland via the auto-dial unit, to the UK OAC at Heathrow and, via GTS, to the OACs in the other countries participating in the pilot project.

ASSESSMENT OF THE UTILITY OF THE COST IMAGE

Assessments of the utility of the COST image—radar plus satellite—began at the Heathrow OAC in April 1985. From the assessments made for which completed forms were available, the results given in Table 2.4 were obtained, using the following classification:

A = very useful (over and above other available data),
B = of limited use (over and above other available data),
C = nil (no weather),
M = misleading.

The assessments were made at three different levels of consideration; by the senior forecaster, for the general forecast and for the local forecast in the immediate vicinity of Heathrow Airport.

The figures in Table 2.4 are those recorded in the questionnaires returned, with no allowance made for any qualifying remarks such as the presence of anomalous propagation (ANAPROP), system faults or wrong inferences being drawn from the satellite data presented.

The senior forecasters' returns take into account the picture of the whole pilot project area, as do those of the general forecast; they may therefore be added together for the purposes of the pilot project, although it is recognized that the senior forecasters' returns would carry more weight than

Table 2.4. Analysis of questionnaire records

	Total	A	B	C	M
Senior forecaster	425	61	259	68	37
General	1054	211	563	215	65
Local	999	90	564	268	77
Totals	2478	362	1386	551	179

Table 2.5. Analysis of consolidated questionnaire records

	Total	A	B	C	M
Western Europe	1479	272	822	283	102
Local (as before)	999	90	564	268	77
Totals	2478	362	1386	551	179

those of the general forecast. The local forecast looks at the UK and, in particular, the south-east of England, in the vicinity of Heathrow. If the first two categories are combined, Table 2.4 simplifies to Table 2.5. Note, therefore, that on 14 per cent of occasions (or nearly 19 per cent if category C—no weather—is eliminated), category A was returned for the European and local forecasts. However, if only Western Europe is considered, on some 23 per cent of occasions a category A—very useful, over and above other available data—was recorded. Moreover, on nearly 91 per cent of occasions when category C is neglected, the radar was of some help (categories A and B taken together) for all forecasts and, again, if Western Europe only is considered, nearly 92 per cent is obtained.

For the local forecasts in the vicinity of Heathrow, the equivalent figures for category A are 9 per cent for all occasions, just over 12 per cent when category C is neglected, and 65.5 per cent and 89.5 per cent for categories A and B combined. On the other hand, on just over 7 per cent of occasions the radar was considered to be misleading (or just over 9 per cent if category C is neglected) for all forecasts. The equivalent figures for Western Europe forecasts where the radar was considered to be misleading are <7 per cent and 8.5 per cent. The figures for local forecasts are: 7.7 per cent and 10.5 per cent.

The reasons why the radar data were considered to be misleading were usually attributed to ANAPROP. Moreover, if one of the radars in the network was out of service for any reason, the other radars in the network tried to cover the gap and were, therefore, working at extreme range with a consequent diminution in accuracy of the COST image. This sometimes caused a category M mark to be given because the information provided by the image was at variance with information gained from other meteorological sources.

It is also interesting to observe that the frequency with which category A marks have been awarded has increased as the evaluation has progressed. A few mistakes of interpretation were made initially. For example, it appears that cloud cover, as represented by the satellite image where no radar data were available, was sometimes mistaken as an indication of precipitation— an all too easy mistake to make until the operator becomes familiar with the system. When this apparent precipitation was not confirmed by other meteorological observations, the COST image was, in consequence, marked down. Such errors of interpretation appear now to be non-existent.

After the analysis was completed a further batch of assessment forms arrived. Cursory examination of these shows that the trend towards up-grading the value of the COST image has continued and only on very few occasions indeed was the 'misleading' mark given. When it was, it was usually due to a radar or system fault. Forecasters in Switzerland have also been examining the utility of the COST image, but for a shorter period. A similar pattern of assessment appears to be unfolding, i.e. initial distrust of the image and a wish for the early inclusion of the French data (this has now been achieved). But, with experience, a more favourable reaction is now discernible.

THE COST PLANNED INTEGRATED NETWORK FOR WESTERN EUROPE

Figure 2.4 shows the twelve radars with digital output which were operational in August 1981 within the COST countries participating in the project. Of these, half were operated independently and only two networks were operational and producing composited radar images at that date—the United Kingdom with four radars in its network and Switzerland with two. The independent national network plans for radar installations with digital output by 1991 are also shown in Figure 2.4.

Figure 2.5 shows the planned COST network which, having taken into account and rationalized the national plans has, it is suggested, produced a more cost-effective network. While the plan may be thought to be totally over-ambitious because it contains 127 weather radars in the complete network, it is salutary to note that no less than 83 radars are already scheduled to be commissioned in the next decade. Spain, for example, from having no plan for a national network at all, has now adopted the COST plan and let a contract for eleven radars to be installed in the next 2–3 years. The UK, too, has issued contract documents for the next six radars to be commissioned. These will be in addition to the five presently in the oper-ational network. Finland, France, Germany, Italy and Sweden are similarly building up their networks, and it is confidently predicted that a small, but growing, integrated, operational network will become a reality over the next 5 years.

Figure 2.4. COST 72. Operational weather picture with digital outputs

POTENTIAL BENEFITS OF THE NETWORK

The primary beneficiaries will, of course, be the meteorological services of the countries taking part in the COST project. Hydrological services will

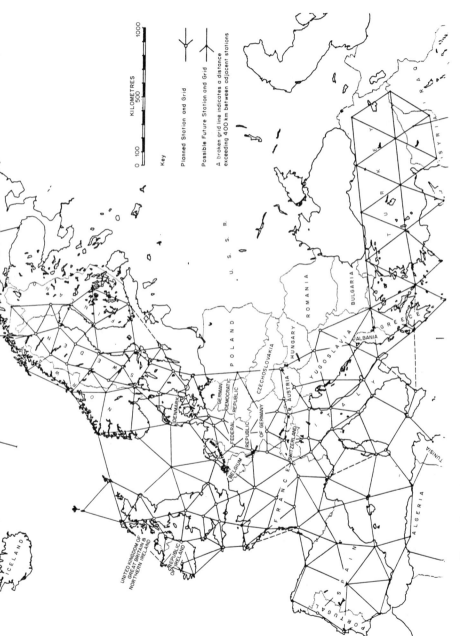

Figure 2.5. COST 72. Possible integrated weather radar network with digital outputs

also benefit eventually, in some cases, as primary data users, but probably in a secondary role initially. This is because they will receive processed data from the meteorological services which can be used, *inter alia*, as input into hydrological models enabling streamflow to be predicted and, when necessary, will enable more precise flood warnings to be issued.

There is a broad spectrum of secondary users—air traffic control, airport operations, transport (road, rail and sea), agriculture, power generation (particularly hydroelectric power generation), and leisure activities (yachting, mountaineering and speleology, for example) who will be able to benefit either from the data from individual radars or from the integrated network where longer-term forecasts are needed.

CONCLUDING REMARKS

COST 72, it can be argued, is an example of European countries working together at their best. Throughout the 5 years of planning, looking at the data that should be exchanged, deciding formats, protocols and methods of data transmission, each country has been responsible for its own costs. Each has made a valuable contribution and has exhibited a high degree of co-operation. That the project has been successful in achieving the objectives set some 7 years ago is without doubt.

Because of this success the formulation of the next phase, the implementation of a pre-operational network, is well advanced and is expected to follow on from the conclusion of the present planning phase with very little delay.

ACKNOWLEDGEMENTS

The help of many people in the participating countries who have contributed to the success of the COST 72 project, and thus to this chapter, is gratefully acknowledged. Special mention must be made of Mr C. G. Collier of the UK Meteorological Office and the Chief Meteorological Officer at Heathrow Airport. The former has made a major contribution to many aspects of the COST 72 project and the latter agreed that his staff, though hard-pressed, would complete the questionnaires to assess the utility of the COST image. The contribution towards the assessment of the COST image made by all the staff of the Meteorological Service is gratefully acknowledged. Figure 2.1 and Table 2.1 are Crown copyright reserved and are reproduced by permission of Her Majesty's Stationery Office. Finally, the content of this paper is derived from work carried out by the author in his role as Consultant to the European Commission on COST 72. Their permission to use the information gained in it is gratefully acknowledged.

REFERENCES

Collier, C. G. and Fair, C. A. (1985). The COST 72 pilot project. WMO Commission for Instruments and Methods of Observing. 8–12 July 1985, Ottawa. *WMO Report No. 22.* 169–73.

James, P. J. (1981). Radar-site signal processing and control in the UK Meteorological Office short period weather forecasting pilot project. *Proc. Seminar/Workshop on Weather Radar, ECMWF Reading,* 9–10 March 1981, pp. 239–44.

Negri, A. J. and Adler, R. F. (1981). Relation of satellite-based thunderstorm intensity to radar estimated rainfall. *J. Appl. Met.* **20**, 288–300.

Newsome, D. H. (1981). Report on user requirements. *Proc. Seminar/Workshop on Weather Radar, ECMWF Reading,* 9–10 March 1981, pp. 5–15.

Weather Radar and Flood Forecasting
Edited by V.K. Collinge and C. Kirby
© 1987 John Wiley & Sons Ltd.

CHAPTER 3

The FRONTIERS Project

G. P. SARGENT

INTRODUCTION

FRONTIERS is an acronym: *F*orecasting, *R*ainy, *O*ptimized using, *N*ew, *T*echniques of, *I*nteractively, *E*nhanced, *R*adar and, *S*atellite data.

The FRONTIERS project was conceived in the late 1970s when precipitation data from a network of radars became available. Methods had already been developed for producing short-period forecasts of precipitation areas and intensities by simple linear advection but the usefulness of radar network data was often impaired by errors, many of which were meteorological in origin and difficult to correct objectively. There was also a need to exercise effective quality control and analysis in real time. Moreover, although the radar network and Meteosat satellite information was useful separately, the limited coverage of the radar data was a major restriction that could largely be removed if ways were developed of combining radar and satellite data to provide analyses of precipitation over an area rather larger than the radar network.

The development of the FRONTIERS system has been a joint project between the Meteorological Office and Logica Ltd. The main development work was done at the former Meteorological Office Radar Research Laboratory at Malvern, Worcs. and is now being continued within the Satellite Meteorology Branch of the Meteorological Office at Bracknell, Berks. It is intended during the summer of 1986 to bring the FRONTIERS system into operational use within the Central Forecasting Office (CFO) of the Meteorological Office, and to this end a duplicate system has been procured. The original system will be used for further development and to act as a back-up for the operational system.

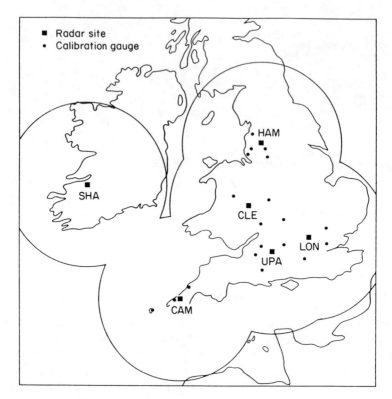

Figure 3.1. The UK weather radar network: CAM—Camborne, UPA—Upavon, CLE—Clee Hill, HAM—Hameldon Hill, LON—Chenies (London), SHA—Shannon

DATA USED

Radar data

The radar data are a composite of data received from up to six weather radars situated as shown in Figure 3.1. Each radar has an on-site minicomputer that pre-processes the radar data and transmits them to the centrally located network computer at Bracknell. The network computer combines all the data received and transmits the resultant composite picture to the users. Associated with each radar are from three to five calibration rain gauges which provide data for real-time calibration of the radar data. The on-site computers also apply some objective corrections to the precipitation data, such as clutter cancellation, interpolation of data where the radar beam is occulted by high ground and corrections to allow for range attenuation of

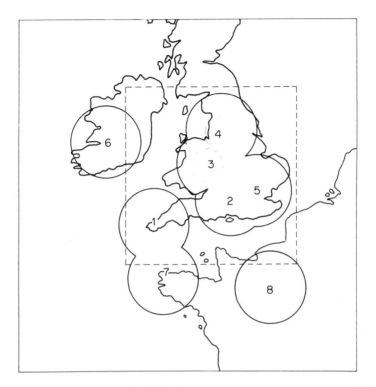

Figure 3.2. The FRONTIERS area: 1 to 6 are the current UK network radars; 7, 8 are French radars it is hoped to include within the network. The dashed square is the area covered by the current JASMIN display system

the radar beam. The composite pictures that are produced have a spatial resolution of 5 km and a temporal resolution of 15 min.

Satellite data

The satellite data are received from the Meteosat satellite by a Primary Data User System (PDUS) located at Bracknell. The PDUS computer processes the data and transmits them to the FRONTIERS system every $\frac{1}{2}$ h. Three types of image are received from the satellite:

1. infra-red
2. visible
3. water vapour.

Within the FRONTIERS system the satellite data are displayed over the area shown in Figure 3.2 with a spatial resolution of nominally 5 km square (in the latitude of the UK it is more like 6 km × 8 km) and a temporal resolution of 30 min (60 min for water vapour during daylight when visible images are available).

THE SYSTEM HARDWARE

The heart of the system is two DEC VAX 11/750 minicomputers with three fixed discs, two removable discs and two magnetic tape drives. One of the two minicomputers is the operational machine and the other is the research and development machine. The system is so configured that in the event of a failure in any part of the operational machine the appropriate parts of the R&D machine can be rapidly substituted and there is minimal loss of data. The minicomputer receives data from the radar network computer and the satellite reception computer via high-speed interfaces. Image data are passed from the VAX to two RAMTEK graphics devices which support two colour monitors and a joystick.

The forecaster operates the system from a workstation (see Figure 3.3), which comprises two colour monitors (1,2), two visual display units (VDUs) (3,7), a datatablet (8), and a joystick (5). One of the colour monitors, the working monitor and the two VDUs are fitted with touchscreens. These enable the forecaster to interact with the system by selecting menu options, drawing areas on the images, etc., by touching the screens with his finger. Thus keyboard entry is kept to a minimum. The datatablet (8) with its 'mouse' (9) enables rather more precise drawing to be achieved than is possible via the touchscreen and the human finger. Provision has been made for the forecaster to choose to use keyboards (4,6) rather than the touchscreens for menu selection if he wishes. (It has been found that some forecasters always use the keyboards while others always use the touchscreens.) Images, both actual and forecast, are archived on to magnetic tape and can then be examined off-line for assessment and training.

THE SYSTEM IN OPERATIONAL USE

The system is menu-driven, i.e. the forecaster is presented with a series of menu options that he may select. Because the Meteosat satellite data are received at $\frac{1}{2}$ hourly intervals the system works to a $\frac{1}{2}$ h cycle. In each $\frac{1}{2}$ h the forecaster is led through three stages:

(a) the radar analysis
(b) the satellite analysis
(c) the forecast.

If the forecaster falls behind a predetermined schedule the system goes into an automatic mode and applies default corrections. This prevents a 'log jam'

Figure 3.3.   The FRONTIERS workstation. 1, Display colour monitor; 2, working colour monitor; 3, LH menu VDU; 4, LH menu keyboard; 5, joystick; 6, RH menu keyboard; 7, RH menu VDU, 8, Bitpad; 9, Bitpad 'mouse'

of data building up. Each of the three stages has a deadline at which automatic mode takes over. When the defaults have been applied the forecaster is allowed to proceed with the next stage.

The radar analysis

The radar data as received from the network computer have been corrected to some extent by the remote site software. Corrections for range attenuation and occultation, as well as some clutter cancellation and interpolation of data in areas where clutter masks the data, have already been applied, together with domain calibration of the data according to the data received from the calibration rain gauges. Before the FRONTIERS forecaster starts to apply subjective corrections, all objective corrections apart from the

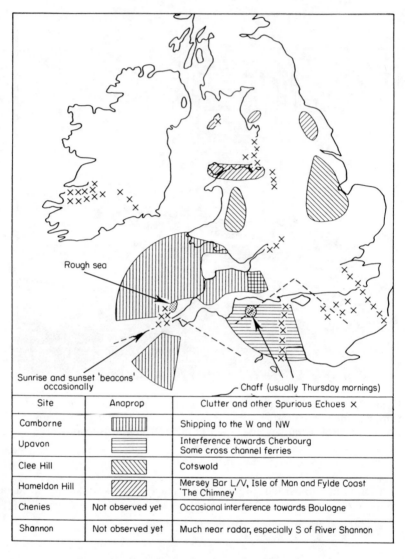

Site	Anaprop	Clutter and other Spurious Echoes ✗
Camborne	⦀⦀⦀⦀⦀	Shipping to the W and NW
Upavon	▤	Interference towards Cherbourg Some cross channel ferries
Clee Hill	▨	Cotswold
Hameldon Hill	▨	Mersey Bar L/V, Isle of Man and Fylde Coast 'The Chimney'
Chenies	Not observed yet	Occasional interference towards Boulogne
Shannon	Not observed yet	Much near radar, especially S of River Shannon

Figure 3.4.   Locations of known spurious echoes

occultation corrections and clutter cancellation are removed from the data automatically.

The radar data are available to the forecaster at about HH + 7 minutes, where HH is the hour or the $\frac{1}{2}$ h, and the forecaster is led through a series of steps each of which addresses a particular problem with the radar data. One of the more important steps is the removal of spurious echoes, that is echoes returned from other than precipitation (see Figure 3.4). These echoes can be from a variety of causes such as clutter, anomalous propagation, etc. A characteristic of such spurious echoes is that they are either stationary or move erratically, whereas precipitation echoes normally move in a coherent manner. The forecaster can replay a time sequence of radar images and determine which echoes are due to precipitation and which are spurious. Rain-gauge data and satellite imagery can also be used to decide if precipitation is likely in a given area.

The spurious echoes can then be deleted by drawing round them and selecting the delete option for all data inside the drawn area. Alternatively, individual pixels of data may be deleted by selecting the option to delete pixels and touching the pixels it is required to delete. In order to facilitate the removal of small areas of data it is possible to zoom and pan the image. Thus the forecaster can 'home in' on an area of interest with a magnification of up to 12×. A useful facility is the 'last times' option. This displays a translucent overlay showing the area where echoes were deleted in the previous cycle. Selecting 'implement' will then remove all echoes under the overlay. When all the spurious echoes have been removed from the displayed image, selecting 'Implement' removes the deleted pixels from the dataset and the system proceeds to the next step.

Other steps in the Radar Analysis stage are corrections for Bright Band errors, Definition of Rainfall Types, Radar Calibration. Adjustments and Orographic Enhancement corrections.

The Bright Band step allows the forecaster to apply corrections for the enhanced returns from melting snowflakes, which appear to the radar as very large raindrops. He can either use the corrections that have been determined objectively by the remote site software, or he can define his own model of the bright band profile and so tailor the corrections more closely to the observed precipitation.

The Definition of Rainfall Types step gives the forecaster the opportunity to define different areas of the radar image as representing different types of rainfall—i.e. shallow, moderate or deep and either layer or convective. The purpose of this classification is so that the right range-dependent corrections are applied to the radar data. With shallow layer cloud, for instance, the radar beam will partially overshoot the rain at relatively near ranges and thus the beam will not be fully filled by precipitation, giving a reduced return, whereas deep convective rain will probably fill the beam right out to the limit of the radar coverage. Layer and convective rain are also composed of different precipitation drop sizes, and the classification allows the optimum value of the relationship between reflectivity of the precipitation particles and the derived rain rate to be used.

The Radar Calibration step allows the forecaster to change the calibration of a radar if it is thought that it is not showing the correct precipitation intensities. The evidence for such errors is found in the ratios of radar to rain-gauge rates, or in obvious inconsistencies between one radar and its neighbour.

In the Orographic Enhancement step the forecaster can make allowance for low-level growth of precipitation over high ground. He defines the low-level airflow parameters of wind speed, wind direction and humidity, and the system applies corrections contained in a statistically constructed set of look-up tables.

When all the corrections have been applied the forecaster has produced what he considers to be his best estimate of the precipitation field within the area covered by the radar network. If the forecaster has not completed the radar analysis by HH + 24 min the system completes it automatically. Plate 1 shows a typical cleaned-up radar image.

The satellite analysis

The satellite data received from Meteosat are available to the forecaster at about HH + 17 min provided he has completed the radar analysis. In this stage the satellite imagery is first navigated (if necessary and/or possible) by moving the displayed coastline overlay over the image until it matches any coastline features visible in the image. A space-view image is used, the forecaster 'zooming in' on an area where coastline features are apparent in the data (see Plate 2). If there is a complete cloud cover then navigation is not possible. At night, when only the infra-red imagery is available, there may not be sufficient temperature contrast between the land and sea to enable the coastline position to be determined, although manipulation of the slicing of the image can be used to enhance the contrast. If navigation of the image is not positively possible the image is left as it is. Plate 3 shows an infra-red image that has been registered and reprojected on to a national grid projection. False colour is used to show the different cloud top temperatures—see the scale of temperature levels at the top left of the image. The temperature range is approximately from +30 to −80 °C.

Having navigated the image, the next step is to produce an estimate of the precipitation field over the whole of the FRONTIERS area. This is done in one of two ways:

1. A correlation table is constructed matching precipitation intensities as seen by the radar with reflectance and radiance values of the visible and infra-red data respectively at the same point. This table is then used to calculate the probability of precipitation at each pixel point in the image. Where the probability exceeds a given threshold, precipitation is indicated. Best results are obtained when there are both visible and infra-red data and when there are large areas of rain within the radar network area (see Plate 4). Note that this image is displayed in a single colour—

namely the highest intensity. This is to distinguish satellite-derived data from radar-derived data in later stages.

2. Either the visible or the infra-red image can be sliced to represent precipitation. Often, but by no means always, precipitation is associated with the coldest and brightest cloud tops. Thus by slicing the reflectance or radiance levels to leave just the brightest or coldest, a representation of the precipitation field can be produced that is reasonably realistic. The forecaster can inlay the radar network data into the satellite image while slicing, and so produce as good continuity at the radar network boundary as possible.

The next step in the satellite analysis stage enables the forecaster to modify the rainfall field estimate outside the radar data boundary. Thus if it is deemed appropriate, areas of data can be removed (in the same manner as spurious echoes are removed from radar data) or areas of colour representing precipitation can be inserted by the forecaster. This latter is sometimes desirable where a continuous band of rain is present in the radar data (and/ or surface observations) but which has not been continued by the radar/ satellite correlation or the satellite slicing technique. In practice it is found to be very difficult to insert authentic-looking precipitation areas without spending a great deal of time.

If the forecaster has not completed the satellite analysis by HH + 30 min the system will complete it automatically. When all modifications have been made the forecaster has his best estimate of the precipitation field over the full FRONTIERS area. This field now forms the basis of the 6 h forecast.

The forecast

After completing the radar and satellite analysis stages the system reduces the data from 5 km resolution to 20 km resolution, and any orographic enhancements that were applied in the radar analysis stage are removed (see Plate 5). The reduction in resolution is made in the interests of speed of computation of the forecast images. The orographic enhancements are by their nature stationary, and are not required to be advected by the forecast velocities.

When the data have been reduced and disenhanced the resulting precipitation field is displayed as a single cluster (see Plate 6). The forecaster must now divide this cluster into sub-clusters which consist of areas of precipitation moving at the same velocity. He is helped in this by replaying time sequences of radar and/or satellite data. Sub-clusters are defined simply by drawing round the appropriate area on the touchscreen and selecting 'Divide cluster'.

Each sub-cluster now has to be given an appropriate velocity which is done in one of three ways:

1. By using the joystick. Selecting this option displays a compass rose on the screen and the forecaster uses the joystick to move a cursor to the point on the rose representing the desired vector.
2. By measuring the displacement of a singular feature in the data (radar or satellite) between two successive images. When this option is selected the previous cycle's image is displayed and the forecaster is prompted to touch the feature of interest. The current image is then displayed and the forecaster is again prompted to touch the same feature. The system then measures the displacement and computes the velocity.
3. By using the Lagrangian replay facility. With this method the forecaster sets up a short reply sequence of (say) five or six images and then touches the same singular feature in the first and last images in the sequence. As in the previous method the system computes a velocity which is used as a first estimate. The forecaster then selects the Lagrangian replay option, whereupon the sequence is replayed continuously, applying the first estimate velocity to the data. If the first estimate is a good one the selected feature will appear to be almost stationary within the frame. The forecaster then uses the joystick to make fine adjustments until the feature does appear to be stationary. The system then computes the new velocity.

These methods are used as required to define velocities for all the sub-clusters, and the forecaster assigns the computed velocities to the sub-clusters. In this example only one cluster is defined at this stage; further subdivision is done later. The system then computes the six forecast images by moving each sub-cluster at the velocity assigned to it. A replay of a sequence of the seven previous cycles' actual images, the current actual image followed by the six forecast images is presented to the forecaster. who then decides if he wishes to make any modifications to the forecast parameters. When he is satisfied with the forecast he can re-apply the same orographic enhancements as were applied in the radar analysis stage or define new enhancements,

The forecaster has until HH + 36 min to complete the forecast stage, otherwise the system will take over and complete it automatically. This concludes the Forecast stage and the cycle is completed by the system automatically applying the same corrections to the 15-min radar data as were applied by the forecaster to the $\frac{1}{2}$ hour data. It will be seen that if the forecaster falls behind schedule and allows the system to go into automatic mode, he will not be able to start the next cycle before about HH + 42 min, putting him under time pressure right from the start of the next cycle.

THE SYSTEM IN OFFLINE USE

The R&D system has the facility to be run in a so-called 'Experimental' mode. This allows data previously archived onto magnetic tape to be loaded and operated upon in the same manner as in operational mode. The forecaster has the option to work with the same time constraints as in operational mode, being led through the steps in sequence, or in 'free-form' with no time constraints and with the ability to select the steps which he requires in any order. This latter mode is invaluable for training new operators and examining cases which proved 'difficult' in real time, allowing the forecaster to spend as much time as is required to study the problems and build up his expertise. System developers can test new algorithms, etc. on real data, either at leisure or within the operational time constraints.

FUTURE DEVELOPMENTS

The system is at present (1986) still in the development stage but it is planned to implement the operational system in the near future in the Central Forecasting Office (CFO). Forecasters in CFO will operate this system as part of their normal duties, disseminating the quality-controlled radar composite images to recipients of Jasmin data. The R&D system will be used for further development and will also act as a back-up for the operational system. In the event of a computer failure in the operational system, it will be possible to switch to the development computer and preserve the database, allowing operation to continue with no more than one cycle's loss of data.

Development of the system will continue with particular emphasis on improving the bright band correction procedures and the radar/satellite correlation algorithms. When the extended area precipitation field estimates are deemed to be sufficiently reliable and useful, these will be disseminated to users. As new radars are added to the existing network their data will be included in the FRONTIERS database. This will add to the workload on the forecaster, making it necessary to streamline some of the processes to avoid unnecessary duplication of tasks and shorten the time taken to implement each step.

Much work remains to be done to improve the accuracy and usefulness of the forecast images. The present linear extrapolation of current precipitation areas leaves a lot to be desired, especially in situations with rain bands rotating about a moving centre. It is impossible with current techniques to represent such movement realistically, and thus it is likely to be some time before forecast data are disseminated to users. As the system is developed, other products, possibly tailored to suit individual users, will become available.

CONCLUSION

The FRONTIERS system, which has been developed over the past few years, is capable of combining the timeliness and wide coverage of radar and satellite data with the skill and experience of the human forecaster to produce, in near real time, realistic estimates of precipitation fields over the radar network area and to provide useful guidance over a much wider area. The interaction of the forecaster with the data allows anomalous or spurious effects to be recognized and corrected before the data are disseminated to the users.

Further development of the system will lead to improved precipitation field estimates outside the radar network area, and to the provision of short-period quantitative forecasts of precipitation intensity. Other products, as yet undefined, will also become available and there is scope for dialogue between the users and the Meteorological Office to determine what products are desired and feasible.

Weather Radar and Flood Forecasting
Edited by V.K. Collinge and C. Kirby
© 1987 John Wiley & Sons Ltd.

CHAPTER 4

A System for Processing Radar and Gauge Rainfall Data

B. R. MAY

INTRODUCTION

PARAGON[1] is a radar and gauge rainfall data processing and storage system developed by the Advisory Services Branch (Met O 3) of the UK Meteorological Office, Bracknell. Its purpose is to provide rainfall data for research purposes and to support an advisory service for external customers. PARAGON consists of two main routine processing sub-systems for producing daily rainfall totals from radar observations adjusted by gauge observations; off-line (non real time) and on-line (near real time). The daily rainfall off-line system became operational in January 1986, and the on-line system became operational at the end of 1986. A third sub-system, for producing sub-daily rainfall data on a non-routine basis, is also planned. This chapter briefly describes the PARAGON system and progress on its implementation.

RADAR DATA

The five radar installations on the UK mainland shown in Figure 1.2 provide observations of rainfall averaged over 5 km × 5 km squares every 5 min. After being given an on-site calibration using observations from dedicated gauges and the removal of permanent echoes, the radar data are recorded on-site and also observations every 15 min are transmitted to the Meteoro-

---

[1] PARAGON is the acronym for *P*rocessing and *A*rchiving of *RA*dar and *G*auge data *O*ff-line and in *N*ear real time).

logical Office at Bracknell via telephone lines. At present, the data are not corrected to remove the effects of bright band, i.e. the apparently enhanced rainfall occurring when the radar beam intercepts melting snow. The recorded data are integrated by the Instruments Branch at the Meteorological Office into GMT hourly rainfall totals for each 5 km square within the coverage of each radar (a circle approximately 210 km radius). These hourly totals, which are received by Met O 3 after a delay of some months, are the input to the PARAGON off-line system. The on-line system, which is not yet operational, will use hourly totals derived from the integration of the 15-min observations received in near real time.

THE PARAGON PROCESSING SYSTEM

Further quality control is carried out involving the correction or deletion of areas of occultation and anomalous propagation, and the corrected data are stored in the first of a sequence of data sets shown in the processing flow diagram (Figure 4.1). This data set, XXHOUR, holds single-site, unadjusted (by gauge observations) radar hourly totals. The data are held as 5 km grid point values in this and all other data sets.

Twenty-four successive hourly totals are next summed to give 0900 to 0900 GMT daily totals which are transferred to the next data set DAY RAD UN, of single-site unadjusted radar daily totals.

The next step involves the adjustment of the radar daily totals by the daily totals measured by gauges in the sparse synoptic network (spacing averaging 40 km) and the resulting radar data are put into the data set DAY RAD AD of single-site radar daily totals adjusted by sparse gauges. This adjustment brings the radar observations into agreement with the surface gauge observations and helps the next process, that of compositing the overlapping coverages of neighbouring radars, to provide a single coverage without noticeable boundaries. This composited field is put into the data set DAY COMP, of composited daily adjusted radar totals.

Because of the delay in processing the off-line radar data to this stage, there is now the opportunity of applying a further adjustment of the composited radar data by the observations from the gauges in the much more dense climatological network (spacing averaging 8 km) which are also not available for several months. At the same time areas of the UK (excluding Northern Ireland) without radar coverage are filled in with 5 km grid point values derived from gauge observations alone, and the resulting data are stored in the final data set BEST EST of the combined field of adjusted daily radar totals with gauge fill-in. Gauge-derived values are distinguished from radar-derived values by a negative sign.

The processing of the on-line radar data for more immediate use will proceed exactly as for off-line data except that the adjustment will be by the

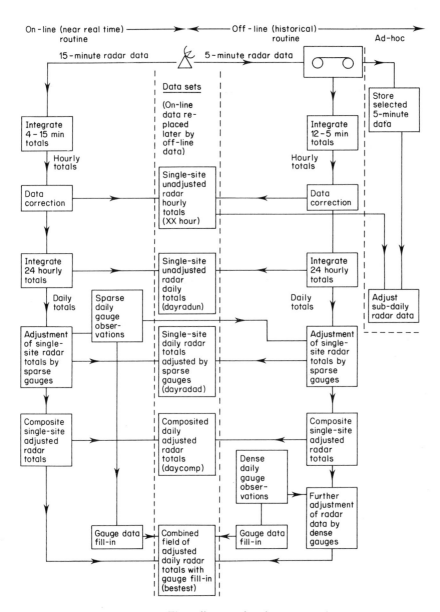

Figure 4.1. Flow diagram for data processing

sparse gauge observations only. The on-line products will be processed and stored within a few hours after 0900 GMT each day, and the on-line contents of all data sets will be replaced later by the more accurate off-line products as they become available.

MINILOG

The elaborate PARAGON system requires a log to indicate the state of the data and the stage of its processing. MINILOG holds such information as data availability (which radars were working), data quality (detection of ANAPROP, gauges used for adjustment), status of data (on-line, off-line) and useful comments on the performance of the radar systems.

AVAILABILITY OF DATA AND INFORMATION

An example of BEST EST daily rainfalls is given in Figure 4.2. Data can be displayed in this way as arrays of grid point totals on page printout or on fiche or 35 mm film using existing software. Procedures for transferring data to magnetic tape will be written according to users' requirements.

The PARAGON data sets are complete from January 1984; data from the Chenies radar only started in January 1985. Data from May 1981 to 1983 have been processed in PARAGON but are less complete and of lower quality than the more recent data.

All hourly totals are stored routinely in the PARAGON data set XXHOUR, but for certain events and periods of special interest, 5-minute totals for 5 km squares (and 2 km squares within 75 km of each radar) are retained by Met O 16 outside of PARAGON. None of these data are adjusted by gauge observations, and because of the large volume of this sub-daily data they will be adjusted only as and when required for special purposes.

Regular reports, entitled *Radar Data Users' Quarterly Liaison Report*, are produced, and consist of two sections. The first contains information regarding the radar site hardware and software, and the second contains statistics of the accuracy of radar data from colocated radar and gauge observations of rainfall amount and occurrence.

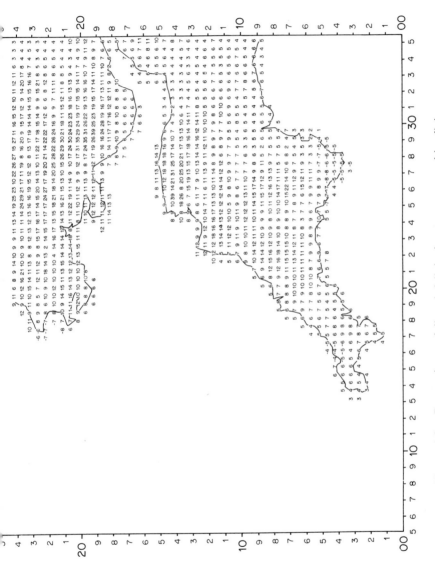

Figure 4.2.   Best-estimate daily rainfall (radar + gauge combined) in mm

Part II
OPERATIONAL EXPERIENCE OF WEATHER RADARS

The most recent operational radar experience in the UK comes from the North West Weather Radar Project, and Chapters 5, 6 and 8 discuss the experience in radar performance, accuracy of rainfall measurements and use in flood forecasting. The use of data from the North West Radar and from the UK radar network at the Manchester Regional Forecasting Centre is the subject of Chapter 7.

Weather Radar and Flood Forecasting
Edited by V.K. Collinge and C. Kirby
© 1987 John Wiley & Sons Ltd.

CHAPTER 5

The Establishment and Operation of an Unmanned Weather Radar

G. HILL AND R. B. ROBERTSON

INTRODUCTION

The North-West Radar Project (NWRP) was established in 1977 and led to the installation of the first unmanned weather radar in the United Kingdom. Work within this project has bridged the gap between earlier research, notably the Dee Weather Radar project, and what is now a fully operational system routinely used by the Meteorological Office, North West Water and others in the water industry. The value of weather radar for the measurement and short-term forecasting of rainfall has been convincingly demonstrated.

The North-West Radar Project was funded by a consortium comprising the Meteorological Office; the North West Water Authority; the Water Research Centre; the Central Water Planning Unit and the Ministry of Agriculture, Fisheries and Food. The aims as specified by the Project Steering Committee were as follows:

1. To establish an unmanned operational weather radar station providing precipitation data in real time, fully integrated with the North West Water Authority's Regional Communication Scheme.
2. To obtain experience of the technical performance, reliability and operating costs of the radar system.
3. To utilize the radar-derived data as an additional source of information, integrated with existing and developing forecasting techniques, to give quantitative precipitation forecasts.
4. To develop hydrological forecasting techniques, using radar-derived data, and to incorporate these into the water authority's operational systems.

5. To assess the benefits of data from a weather radar installation in the North West Water Authority's area, including the following applications:
   (a) Flood forecasting (to include a routine for flood-prone areas, and a procedure for flood warning under rare conditions of extreme precipitation).
   (b) Urban run-off management (including worker safety, control of pumps, sluices, marine outfalls and sewerage works).
   (c) River regulation.
   (d) Heavy rainfall warnings.
   (e) Snowfall measurement.

The proposal that the first unmanned operational radar designed specifically for precipitation measurement should be sited in the area of the North West Water Authority was based on the following considerations:

1. A radar sited in central Lancashire would scan a wide range of different types of catchment, covering all operational aspects.
2. The water authority was implementing a Regional Communications Scheme (described in Chapter 8) based on conventional outstations and radio links with central computer control. This would facilitate the operation of the radar and the integration of its calibrating rain gauges. Further, it would provide links with the control centre.
3. The Regional Communications Scheme would provide the opportunity to evaluate the performance of an unmanned weather radar, compared with that of conventional telemetering rain gauges.
4. There would be considerable potential for enhancing the benefits of the water authority's flood warning scheme.
5. There were possibilities of capital savings on the flood warning scheme.

SITE SELECTION

The criteria used in the selection of a site for construction of the weather radar station were:

1. good radar coverage of the areas where quantitative measurements will be most useful;
2. good radar coverage of urban areas, subject to ground clutter cancellation techniques;
3. minimum area affected by permanent echo, because this prevents practical measurement by rainfall;
4. acceptance by local planning authorities;
5. reasonable access and communications.

Figure 5.1.   Radar outstation—Hameldon Hill

Fourteen possible sites selected from the maps of the area were studied using computer procedures with digitized topographical contours at 50 m intervals. The three best sites were then visited, and the short list reduced to two: these were Hameldon Hill and Squires Gate, the former being a high-level site which would also cover much of the Yorkshire Water Authority area and the latter a low-level site near Blackpool. The final choice, because of the better coverage offered, was Hameldon Hill. Figure 5.1 shows the radar outstation, Figure 5.2 shows the position of the site, and Figure 5.3 shows the horizon as seen from Hameldon Hill.

MEASUREMENT OF RAINFALL

Weather radar measures the reflected energy returned from raindrops, hail or snow and converts these measurements into estimates of surface precipitation. There is an empirical relationship between radar reflectivity and the rate of precipitation but many variables affect the values in this relationship, as discussed later in this chapter.

In order to see rain near to the ground and out to as long a range as possible, a low beam elevation is required, but then hills and tall structures can interfere with the beam (Figure 5.2). At Hameldon Hill a series of four

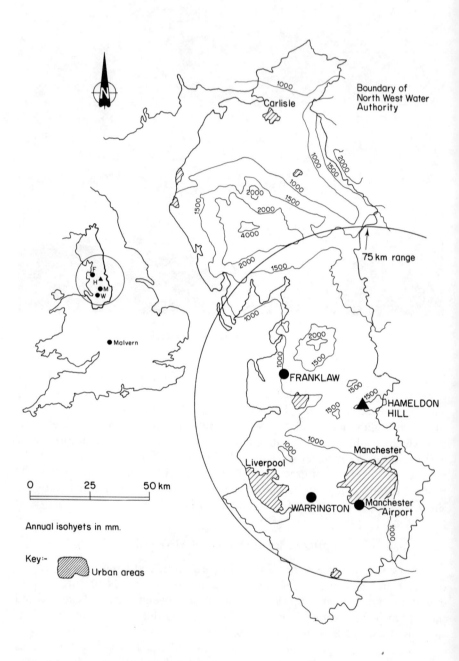

Figure 5.2.   Site of North-West Radar

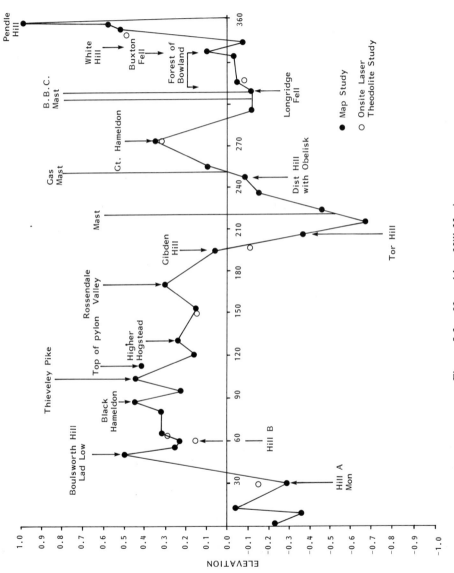

Figure 5.3. Hameldon Hill Horizon

Weather radar and flood forecasting

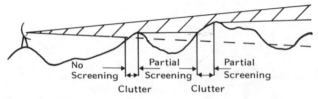

Figure 5.4. Screening of the radar beam and permanent
echo (clutter) in a hilly region.

elevations (0.5°, 1.5°, 2.5° and 4° above the horizontal) are used; the lowest
two elevations being used for estimation of surface rainfall. The higher
elevations are used for bright band detection and other studies.

The radar aerial rotates continuously about a vertical axis at a rate of
approximately one revolution per minute. Each elevation is scanned in turn
and three scans of 'surface rainfall' are available every 15 min the 1.5° beam
being used within 24 km of the radar and the 0.5° beam beyond this range.

Adjustments to raw radar data

There are several corrections which have to be made to the raw radar data:

1. Echoes from permanent objects—ground clutter—are removed by 'sub-
   tracting' the signals received on a dry day from the signals received on
   a wet day.
2. Where the beam is partially blocked by hills, etc. an adjustment is made
   to allow for the reduced beam width beyond the obstruction as shown in
   Figure 5.4.
3. When the radar beam passes through rain the radar signal decreases, as
   does the return echo from rain at a further range. A correction is made
   for this attenuation.
4. Quite apart from any attenuation due to the signal passing through rain,
   increasing distance causes the signal to decrease. This is corrected in
   accordance with an inverse square law.
5. At long range, due to the curvature of the earth and the elevation and
   spread of the radar beam, a significant proportion of the beam is likely
   to be above the rain and an appropriate correction is applied (Figure
   5.5).

Changes in the atmospheric conditions can, from time to time, cause the
radar beam to be bent down towards the earth's surface, giving an apparent
increase in clutter. This is known as anomalous propagation and cannot be
corrected automatically.

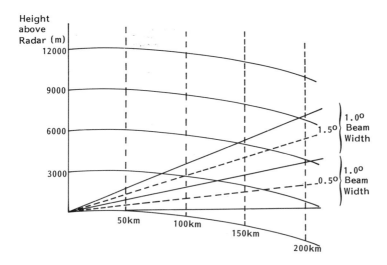

Figure 5.5. Radar beam height v. range for a 1° beamwidth radar at
0.5° and 1.5° elevations

In addition to the above corrections the computed rainfall intensity is compared with data from five telemetering raingauges through the RCS system (Figure 8.1) and adjusted in real time. The techniques that have been developed to make the adjustment using 'ground truth' are fully described in Chapter 6. This adjustment, together with all the above corrections, except anomalous propagation, is carried out automatically by the radar computer.

Radar output

The radar control computer fixes the position of any observation by using polar co-ordinates and then converts the information to Cartesian co-ordinates. Output is available on a 2 km square grid to a radius of 75 km and on a 5 km square grid covering the whole area to a range of 210 km with precipitation information to 208 classes of intensity. The areal rainfall over specified subcatchment areas is calculated every 15 min using these high-resolution data from the three scans obtained during the 15 minutes. Subcatchment values are calculated by the radar computer from the most appropriate 2 km or 5 km grid square values.

Data output is in three forms:

1. 5 km square data in eight levels of rainfall intensity, taken from the last scan in any 15-min period, are used for a colour TV display, and up to 128 subcatchment area rainfall totals are available on a printer. The data

are updated every 15 min and are currently sent to NWW Flood Forecasting Office at Warrington, the Regional Meteorological Office at Manchester Airport, and Yorkshire Water Authority in Leeds.

2. Subcatchment totals are sent every 15 min to the NWW computers at Franklaw for storage and for use in the NWW flood warning system.

3. The fine-resolution data, in 208 levels of rainfall intensity, including data from all four elevation scans, are sent to Met O RRL, Malvern, for archiving and compositing with data from other radars.

For security purposes, approximately 6 days' data are stored on magnetic tape at Hameldon Hill as a back-up in the event of a communications failure.

EQUIPMENT

Radar and computer

The main equipment of the radar station is shown in block diagram form in Figure 5.6. The radar is a Plessey Type 45C, with a beam width of 1° operating within the C band (5.6 cm wavelength) and is intended to be capable of measuring rainfall rates over an intensity range of 0.1 mm/h to 100 mm/h at distances of at least 745 km. Although basically standard equipment, it was modified to allow it to operate in the remote environment of Hameldon Hill. The antenna system has a 3.7 m diameter parabolic dish. It is mounted on the roof of the building and enclosed in a 6.7 m diameter metal space frame radome with panels of reinforced resin-impregnated membrane. Space heaters are provided to prevent icing-up or accumulation of snow.

The original radar-computer interface was supplied by Plessey to meet the customer's specification. The radar signal averaging unit, of proprietary design from Microconsultants Limited, is also in the computer rack. The unit allocates radar video signals to polar cells and averages them prior to computer processing.

The processing system includes a Digital Equipment Corporation (DEC) PDP11/34 computer which initially had 64k words of core store, but has since been increased to 128k words. There is a nine-track magnetic tape recorder for program loading and archiving, synchronous and asynchronous data communications interfaces and a DEC-writer for giving manual instructions and for obtaining a print-out of certain data.

Ancillary and monitoring

A radar monitor, from Magnetic AB (Sweden) is fitted to the transmitter rack to provide continuous monitoring of the transmitter power output and

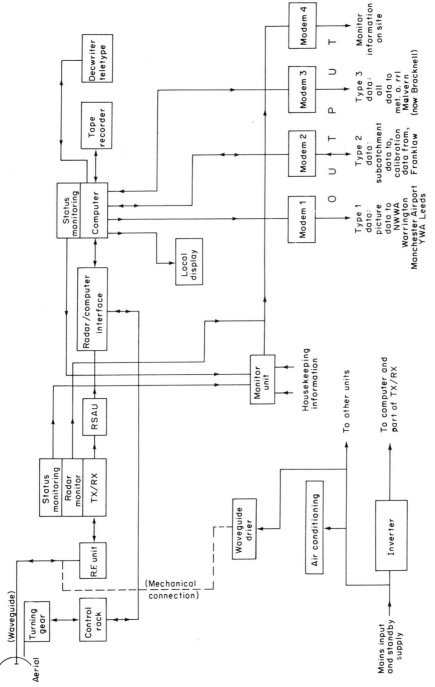

Figure 5.6. Radar and computer equipment

of the receiver noise factor. The fault monitor takes its signals from the radar and also from a computer 'watchdog' for building services (charger fail/fuel tank low/fire alarm/under-or-over room temperature) and from an intruder alarm. It passes signals, indicating which 'fault' has been activated, via the NWW telemetry system to the NWW Control Centre at Franklaw.

Modems pass the data from the computer over the microwave radio link to Franklaw for modelling, and to Warrington both for use there and for onward transmission to Met O RRL Malvern, Manchester Airport Forecast Office and to Yorkshire Water Authority, Leeds. The modem which takes data to Franklaw also brings in calibration rain gauge data.

A Jasmin store with a Digivision colour monitor, identical to that supplied to users, is provided in order that visiting staff may check on picture data. It is also valuable in carrying out system checks where colour patterns are generated to represent the status of certain units.

The system is normally run from the local mains supply, but a standby generator with automatic start-up facility is provided. Although the standby generator comes into use and runs up to speed in about 15–20 sec, this is too long an interruption for the computer and for some radar transmitter circuits. The program would be lost and the radar would need several minutes before transmitting again. Therefore a battery-supported inverter is provided to power the sensitive units. When mains power returns after a break, it automatically takes over and the generator shuts down.

PERFORMANCE

System reliability

From an operational viewpoint the reliability of a 'system' is usually assessed in terms of how often the system performed satisfactorily when it was needed. The user is not particularly concerned with the failures outside periods of operational need. Unfortunately the occurrences of failures of complex systems are extremely difficult to predict, so that design and maintenance objectives have to aim at the highest possible overall reliability with only minimal impact on the frequency of failure during critical operational situations. System design and maintenance arrangements can, however, have a significant effect on time to repair, and hence downtime.

The initially specified mean time between failures was 1000 h for the radar equipment and the overall highest target availability was 99 per cent, i.e. an average downtime of 8 h per month, including repairs and preventive maintenance. It was anticipated that, after experience of unmanned operation for a suitable trial period, the frequency of preventive maintenance visits would be once per month.

The two major constraints which influenced the achievement of high overall reliability of the radar system were the capital cost and the requirement of unmanned operation of the radar station. During the system design stage, duplication of hardware at the radar station was ruled out on economic grounds. However, the provision of a complete set of spare parts (with the exception of aerial spares) was agreed and the most useful location for the spares holding was the radar station. The importance attached to the provision of spares and to the remote monitoring scheme is apparent, since they accounted for 11 per cent of the total cost of the radar station equipment. Likewise, specialized test equipment needed to diagnose critical faults was purchased and held at the radar station so as to be available immediately.

An essential feature of the system is the automatic monitoring scheme which reports critical failures of equipment at the radar station to a remote 24 h manned control centre. This scheme, together with reports from users of the system, enables remedial action to be initiated within 30 min of the occurrence of a fault. The remote monitoring equipment at the radar station provides both pre-fault trend recording and definite equipment failure alarms.

Measurement of reliability

Since users of the TV map data presentation have no need to observe the picture continuously, the collection of statistics on the availability of 'picture data' is impracticable and uneconomic. However, the reception of the high-resolution data is monitored continuously by a microcomputer. Therefore an accessible measure of overall reliability is the availability of these data at Met O RRL. The disadvantages of this criterion as a performance indicator are that data retrieval depends on a BT analogue circuit and that it takes no account of equipment at Manchester Airport and NWW Warrington for the receipt of 'picture data'. However, it was selected as a useful long-term statistic.

The fault monitoring equipment at the radar station enables reasonably accurate specification of the times of occurrence of faults, so by recording resumption of service, maintenance staff can produce a reasonably accurate record of outages due to faults. Downtime for preventive maintenance is recorded as a standard procedure. Overall meantime between failures (MTBF) may provide a worthwhile record but because of the long period of on-site development work there is insufficient data to report MTBF accurately either for the overall system or for any specific part of the hardware. So at this stage, radar station hardware availability is the only other recorded measure of reliability.

Radar computer software failures have been negligible but the loading of updated and improved versions of the software requires the system to be shut down, and this inevitably reduces the overall availability. Software changes are always implemented at times of least operational need.

Table 5.1. Maintainability and availability, Hameldon Hill weather radar, May 1980 to April 1985

	Emergencies					Non-emergencies					Total				
	1980/ 81	1981/ 82	1982/ 83	1983/ 84	1984/ 85	1980/ 81	1981/ 82	1982/ 83	1983/ 84	1984/ 85	1980/ 81	1981/ 82	1982/ 83	1983/ 84	1984/ 85
Number of failures[a]	26	30	12	19	17	14	17	14	1	4	40	47	26	20	21
Routine work (h)	—	—	—	—	—	—	—	—	—	—	81	80	59	56	40
Investigations and modifications (h)	—	—	—	—	—	—	—	—	—	—	136	49	34	6	3
Faults (h)	404	309	85	118	147	32	30	94	13	41	436	339	179	131	188
Total outage (h)	404	309	85	118	147	32	30	94	13	41	653	469	272	193	231
MTBE (h)	312	276	707	451	503	—	—	—	—	—	—	—	—	—	—
MTBF (h)	—	—	—	—	—	—	—	—	—	—	203	176	326	428	407
MTTR (h)	15.5	10.3	7.0	6.2	6.9	2.3	1.8	6.7	13	10	—	—	—	—	—
Availability (%)[b]	—	—	—	—	—	—	—	—	—	—	92.5	94.7	96.9	97.8	97.4
	—	—	—	—	—	—	—	—	—	—	95	96.1	98	98.5	97.9

[a]Faults include all hardware or software failures, whether affecting the supply of data or not.
[b]Availability includes faults/routines/investigations/modifications, etc.

Table 5.2. Breakdown of faults—Hameldon Hill Weather Radar, 1980/85

1980/82[a]	1982/83	1983/84	1984/85	Total	Percentage of total	Fault
6	5	0	0	13	8.4	Computer
23	5	12	12	46	29.8	Computer crashes
19	4	2	2	25	16.2	Radar/antenna
23	4	3	3	39	25.4	Radar ancillaries
16	8	4	4	31	20.2	Services
87	26	20	21	154	100	

[a]Two-year period.

The two measures of reliability are plotted on the graphs in Figure 5.7. The record of hardware availability started in September 1981. The availability of data to Met O RRL at Malvern shows that, on average, there has been a steady improvement since October 1981, with the exception of a 3-month interlude December 1981 and February 1982. During this time an essential hardware component, the radar–computer interface, was upgraded with consequent disruption during installation and testing.

Allowing for an extended period of initial installation and post-installation development of the system, both the mean time between failures of the basic Plessey 45C radar equipment and the overall system reliability are better than predicted and should give encouragement to unmanned operation where appropriate for future weather radar systems.

Performance statistics

A summary of the reliability data is given in Table 5.1, and further details of the areas when faults have occurred are given in Table 5.2. In the early days following installation and commissioning a great deal of investigation was carried out into problems associated with transients and electrical interference. These problems were eventually resolved by the fitting of opto-isolators on the aerial controls and the design of a new radar–computer interface. At the same time many modifications were incorporated into the fault monitor to improve its performance.

It can be seen from Table 5.1 that the availability of the total system, taking account of downtime for routine work and faults, has increased each year except for 1984/85. This increase is mainly a reflection of the reduction in time spent on routines (down from 81 h per year in 1980/81 to 40 h per year in 1984/85) and a reduction in investigations, etc. (down from 136 h

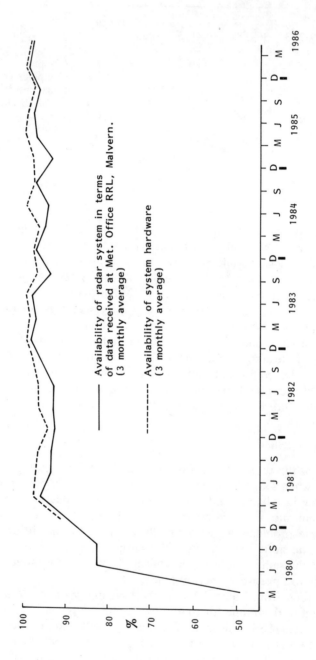

Figure 5.7.   North West Weather Radar—system performance

per year in 1980/81 to 3 h per year in 1984/85). Added to this, of course, is the steady reduction in the number of faults per year spread over the total system.

Availability of the total system naturally increases when all normal servicing work is excluded from the downtime figures (Table 5.1) and in this respect one can see an increase from 95 per cent in 1980/81 to 98.5 per cent in 1983/84 with a small decrease to 97.9 per cent in 1984/85, this latter being due to an increase in the number of computer crashes The weather radar software at Hameldon Hill is to be replaced by a new software package written in Fortran 77, and this will reduce the number of computer crashes, which over the period 1980 to 1985 has accounted for 30 per cent of all faults (Table 5.2).

POTENTIAL SYSTEM IMPROVEMENTS

Since the installation of the weather radar system at Hameldon Hill a second type 45C system using a more advanced design has been installed at Chenies in the lower Chilterns. One of the main differences in the Chenies system is the use of a new Meteorological Office-designed fault monitor which, since its installation in October 1984, has operated very satisfactorily. With some modification it is proposed that a fault monitor to the same design be fitted for a trial at Hameldon Hill. Another important change is the swept gain range at Chenies of 4 km to 100 km; this also is an improvement that could be incorporated at Hameldon Hill to replace the current swept gain unit which compensates for the range attenuation of up to 50 km.

Perhaps the greatest prospect for improving the Hameldon Hill system lies in the data processing area. The software needs to be updated and the now obsolete radar signal averaging unit could be improved by a more reliable array processor. An early proposal to increase reliability by a back-up computer system was not pursued, although consideration may be given to the replacement of the PDP 11/34 by increased processing power.

The incorporation of a confidence indicator for user's displays is still considered to be a useful improvement and, provided that feasibility investigations support the development, will be incorporated as resources permit; likewise a display indicator to inform users of the reasons for loss of data, e.g. fault/servicing, will be incorporated as time and resources permit.

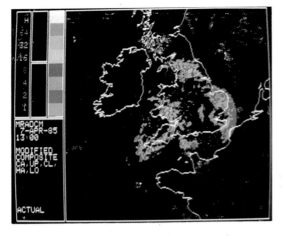

Plate 1. A FRONTIERS
radar image

Plate 2. A space-view visible
satellite image. Note coastline
features apparent in Gulf of
Lyons and south-west France

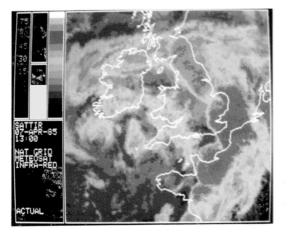

Plate 3. A false colour
infra-red satellite image

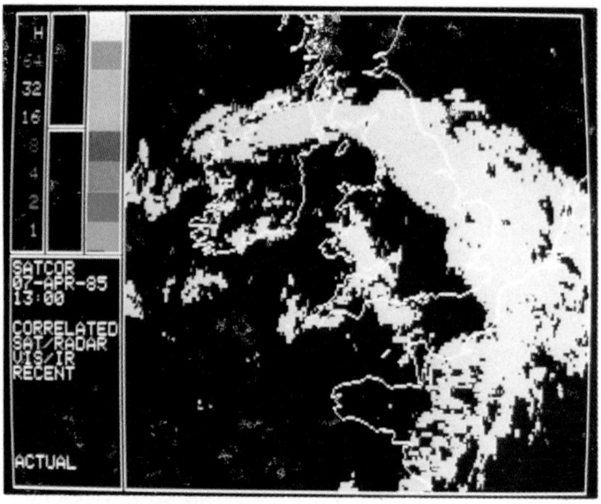

Plate 4.   Satellite derived
precipitation field

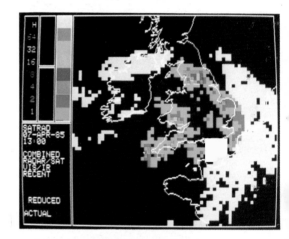

Plate 5.   Reduced data

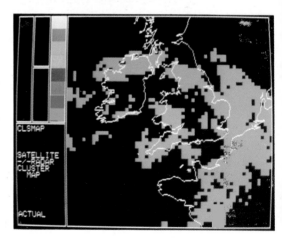

Plate 6.   Cluster map

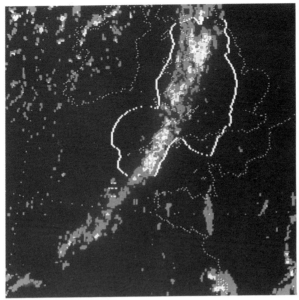

Plate 7. Distribution of rainfall during the passage of a cold front across the British Isles as inferred from radars (within the white outline) and Meteosat (elsewhere)

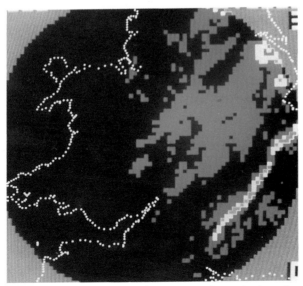

Plate 8. Rainfall distribution as seen by a single radar during the passage of a cold front across England. Areas of light and moderate rain are shown blue. A narrow band of heavy rain associated with line convection at the surface cold front is shown red and pink

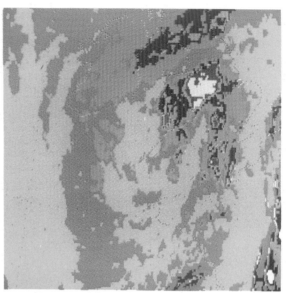

Plate 9. Sub-synoptic comma cloud, with embedded mesoscale convective systems, over England and Wales, as seen in Meteosat infra red imagery. Cold high clouds are shown pink, red and white (white highest). Warm shallow clouds are green and blue. Yellow areas are mainly cloud free. Coastlines are shown in red

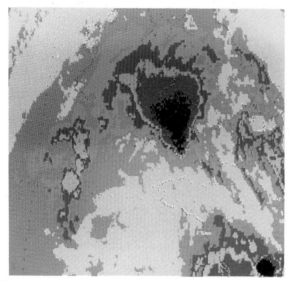

Plate 10. Mesoscale convective system over southwest England as seen in Meteosat infrared imagery. Cold high clouds are shown pink, red and black (black highest). Warm shallow cloud are green and blue. Yellow areas are mainly cloud free. Blue dots show coastline

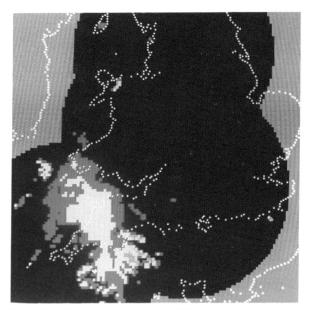

Plate 11. Distribution of rainfall associated with the mesoscale convective system in Plate 10, as observed by the UK weather radar network. Moderate to heavy stratiform rain is shown blue and pink. Heavy convective rain is shown red. Area of coverage is a quarter of that shown in Plate 10 which extends much further south and west of Britain

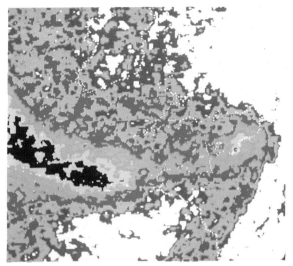

Plate 12. Meteosat water vapour image showing a tongue of dry upper tropospheric air (black, yellow, green and pale blue) corresponding to the dry intrusion shown in Fig 16, 18. Areas of abundant upper tropospheric moisture and cloud are shown red and white

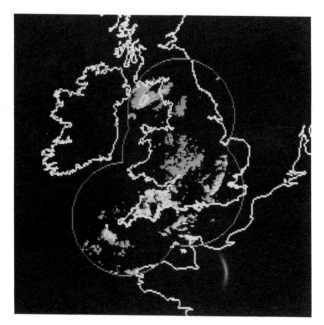

Plate 13.   (details in text)

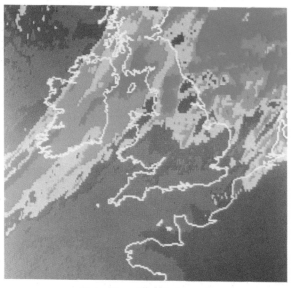

Plate 14.   (details in text)

Plate 15.   (details in text)

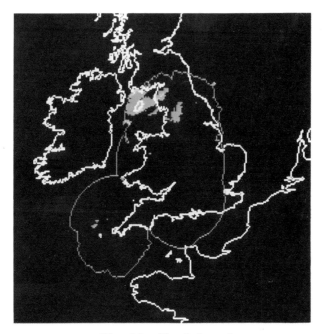

Plate 16.   (details in text)

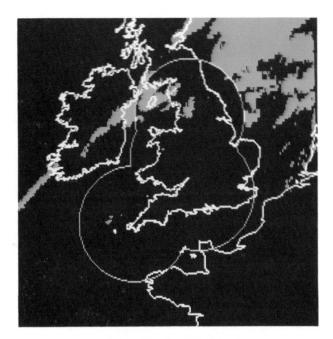

Plate 17.   (details in text)

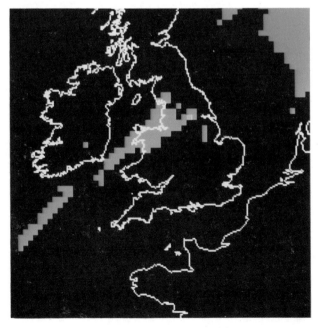

Plate 18.   (details in text)

Weather Radar and Flood Forecasting
Edited by V.K. Collinge and C. Kirby
© 1987 John Wiley & Sons Ltd.

CHAPTER 6

Accuracy of Real-time Radar Measurements

C. G. COLLIER

INTRODUCTION

Radar measures the radio energy returned to the transmitter site after reflection and scattering by raindrops, hailstones or snowflakes. Estimates of rainfall rate R are derived from measurements of the returned energy (radar reflectivity) Z using an empirical relationship of the form.

$$Z = aR^b$$

Many values are possible for both a and b, although b does not vary as much as a. A method of automatically adjusting the factor a in real time has been developed as part of the North-West Radar Project (Collier et al., 1980a), using a computer at the radar site and a number of calibration rain-gauges to improve the accuracy of the derived estimates of rainfall (Collier et al., 1983). This method depends on an understanding of the physical reasons for the variations in the $R : Z$ relationship. These have been found to include the following:

1. Variations in the raindrop size or the presence of hail, snow or melting snow in the radar beam. When the beam intersects the region where snow melts to form rain, the radar reflectivity is enhanced and a 'bright band' is observed.
2. Changes in the intensity of precipitation, both within the radar beam and between the beam and the ground, due to raindrop growth or evaporation. An example of raindrop growth at low levels below the radar beam is

71

the orographic rainfall on the western slopes of the Pennine hills of northern England in suitable atmospheric conditions.
3. Variations in the performance of the radar system.
4. Attenuation of the radar signals due to heavy rainfall along the beam, and to effects of water on the radome.
5. Ground echoes detected as a result of anomalous propagation of the radar beam.

The effects of the first two types of variation are dominant (Browning, 1981), although the others may be important on particular occasions. This chapter examines the accuracy of radar measurements of rainfall made in real time, drawing largely upon the results of the North-West Radar Project (NWRP).

ASSESSMENT FACTOR

To investigate how to allow for variations in the $Z : R$ relationship it was necessary to explore the reasons for the variations. To do this, radar estimates of hourly rainfall were compared with hourly readings from a small number of telemetering raingauges. Hourly values were used for assessment because a real-time procedure must respond to changes as quickly as possible, yet the integration period must not be so short as to introduce large sampling errors. In the present work the radar scans the 'rainfall rate' over each grid square twelve times per hour and the twelve estimates are integrated to form hourly totals. The raingauges used were of the tipping bucket type (0.2 mm per tip) and they were sited in differing topographical situations at various distances and directions from the radar site (Figure 6.1). Harrold *et al.* (1974) showed that a sampling rate of twelve times per hour could, on occasions, introduce errors of up to 10 per cent, and the tipping bucket raingauges of the type used in this study were subject to errors of a similar magnitude.

For assessment purposes the radar estimates in each case relate to a 4 km × 4 km square selected from the standard grid so as to control the relevant raingauge, and the ratio of the radar estimate of rainfall to the raingauge reading is defined as an assessment factor A, i.e.

$$A = \text{radar estimate/raingauge reading}$$

Clearly the raingauge readings will not always represent the true average rainfall over the 4 km square, and discrepancies will arise from the horizontal drift of the rain as it falls from the height of the radar beam to ground level. No attempt has been made explicitly to allow for these relatively small scale effects, as it was felt that larger errors could be introduced in some situations.

Figure 6.1. Topography of North-West England and North Wales

Variability of assessment factor

The Dee Weather Radar Project (DWRP) (Central Water Planning Unit, 1977) investigated the accuracy of hourly rainfall over subcatchments of the River Dee in North Wales, a total area of about 1000 km² (Figure 6.1). Whilst real-time calibration of the radar data over this limited area was achieved, the problems of calibration over the entire 15 000 km² area of quantitative coverage within 75 km of a radar were not addressed in the DWRP. In Figure 6.2(a) values of A, the assessment factor, obtained at the Alwen rain gauge site (DWRP) during the summer of 1976 are plotted against the hourly rainfall intensity recorded by the raingauges; for comparison Figure 6.2(b) shows similar data obtained at the Abbeystead rain- gauge site (NWRP) during 1980. In each case the variability of the assessment factor is very large. This is using only the typical relationship:

$$Z = 200 \, R^{1.6}$$

Joss *et al.* (1970) found the following distinct relationships:

$$\text{Drizzle } Z = 140 \, R^{1.5}$$

$$\text{Widespread rain } Z = 250 \, R^{1.5}$$

$$\text{Thunderstorms } Z = 500 \, R^{1.5}$$

although these relate to an alpine area with quite different rainfall characteristics from those of north-west England. In Figure 6.2(b) cases of showers and thunderstorms are differentiated from cases of widespread frontal rain. These data give mean relationships of $Z = 180 \, R^{1.6}$ for widespread rain and $Z = 240 \, R^{1.6}$ for showers.

The type of rainfall certainly contributes to the variability of the assessment factor, but significant variation also occurs within the frontal rainfall category. Figure 6.3(a) shows the variation of the assessment factor with the 900 m wind speed and direction at the Abbeystead raingauge during frontal rainfall with no bright-band effects present. In this diagram the radial lines represent wind direction and the distance along a radial represents wind speed. Figure 6.3(b) is a sketch map of the local topography around the raingauge site. Figure 6.4 shows a similar diagram and sketch map for the Swineshaw rain-gauge. (The location of raingauges mentioned in this chapter are shown in Figure 6.1).

From the diagrams it will be seen that generally, when wind is from the west so that these gauges are on the windward slopes of the hills, the radar increasingly underestimates the rainfall as wind speed increases, and this indicates the increasing development of orographic rainfall at very low levels,

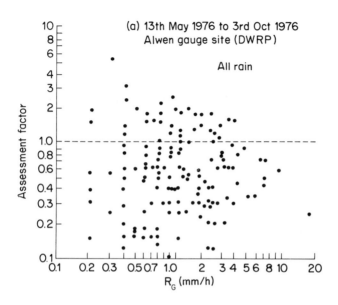

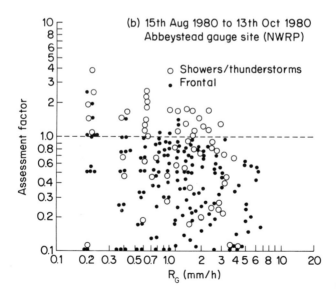

Figure 6.2.   Variation of hourly assessment factor with hourly rain gauge totals

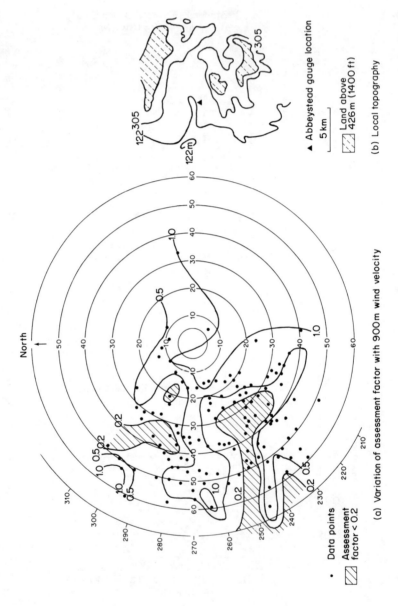

▲ Abbeystead gauge location

⌐ Land above
▨ 426m (1400 ft)

(b) Local topography

(a) Variation of assessment factor with 900m wind velocity

• Data points

▨ Assessment
factor < 0.2

Figure 6.3.   Variation of assessment factor with 900 m wind velocity for Abbeystead rain gauge site

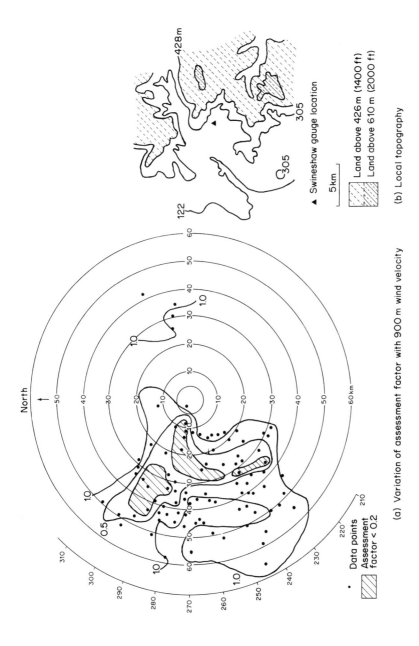

(a) Variation of assessment factor with 900 m wind velocity

(b) Local topography

Figure 6.4.   Variation of assessment factor with 900 m wind velocity for Swineshaw rain gauge site

below the radar beam. However, for the Swineshaw site particularly, this effect reverses as the wind speed increases beyond about 40 knots (20 m s^{-1}). Figure 6.5 shows the vertical structure of rainfall intensity, along a line running south-west from the radar site, for two different south-westerly wind speeds. These diagrams illustrate the wealth of information that can be obtained from the radar using upper beam elevations. The orographic rainfall development at low levels with a medium wind speed is clearly shown in Figure 6.5(a) and the relative absence of this effect at high wind speeds is confirmed by Figure 6.5(b). The large scale structure of northern England and North Wales, shown in Figure 6.1, suggests that this reduction in orographic rainfall at high wind speeds is caused by the influence of the mountains of North Wales. This is referred to later in this chapter as the 'Welsh shadow' effect. There is also some evidence that the mountains of the Lake District have a similar effect at the Abbeystead site during strong north-westerly winds. It would appear that much of the variability in the accuracy of radar estimates of frontal rainfall may result from the interaction between the air flow and the local and larger-scale land form.

Figure 6.6 shows a typical example of how the assessment factor at a site can vary over a period of a few hours. This order of variability—from 10 to 0.1—is entirely due to reasons (1) and (2) discussed in the Introduction. Assessment factors much larger than unity are often evident in showers and in bright-band situations. Values much less than 1.0 are often observed within the warm sectors of depressions (Collier *et al.*, 1983). Variations are rapid in showers, slower in frontal rainfall and bright-band situations, and small in rain falling within strong south-westerly winds in the moist warm sector of a depression (the 'Welsh shadow' effect). Collier *et al.* (1983) have shown that the variations in the assessment factor can be used to distinguish between frontal rainfall, rainfall associated with the Welsh shadow effect, and showery rainfall, as described later in this chapter.

THE CALIBRATION OF RADAR ESTIMATES OF RAINFALL USING RAIN GAUGE DATA

Considerable work has been carried out over the past 30 years or so into the relationship between rainfall and radar reflectivity (for a summary see Batton, 1973) and attempts have been made to calibrate radar estimates of rainfall using raingauge readings (for a review see Wilson and Brandes, 1979). In some cases a small number of raingauges has been used to remove apparent random fluctuations in assessment factors, (Wilson; 1970, Cain and Smith; 1976); an average calibration is applied to the whole area of interest. In other cases a large number of raingauges have been used to define the spatial variations in the assessment factor (Brandes, 1975).

The first method is unlikely to improve the radar accuracy over a large

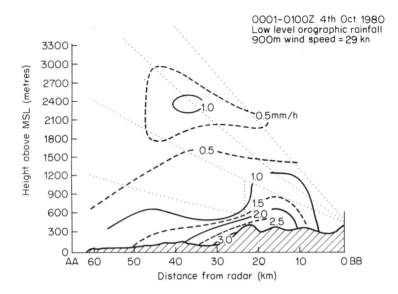

(a) A case of low level rainfall enhancement at medium
southwesterly wind speed

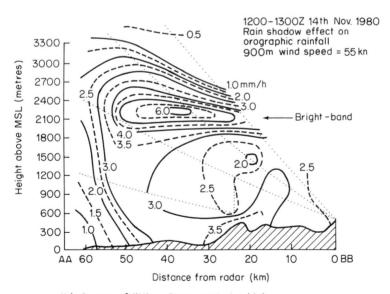

(b) A case of little enhancement at a high
southwesterly wind speed

Figure 6.5.   Wind speed enhancement of rainfall on a cross-section along the line
AA–BB in Figure 6.1

80

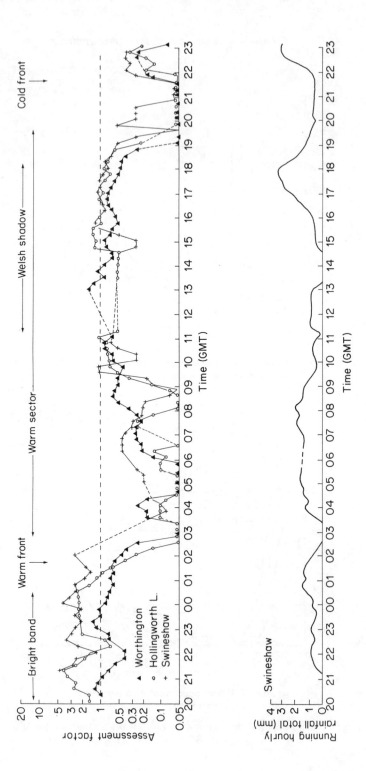

Figure 6.6. A small depression with bright-band effects in front of a surface warm front, 1/2 January 1981

area because of the spatial variations, and the second is unsuitable for real-time operation because a large number of telemetering raingauges would be required. The next section discusses new ways of improving real-time radar estimates of surface rainfall, using data from only a small number of telemetering raingauges. The approach involves consideration of the physical reasons for the variations in the assessment factor.

A new approach to real-time calibration

Having established that, within frontal rainfall, the spatial variations of the assessment factor are related to the topography, appropriate calibration factors can be applied to the radar data when this type of rainfall is identified. Such topographic, or more correctly orographic, effects may be obscured by bright bands, and they are not evident in showery rainfall. It is therefore important to be able to recognize the broad rainfall type if appropriate calibration procedures are to be applied. This is a relatively simple matter for a meteorologist with access to synoptic data, but poses problems for an automatic real-time procedure.

Collier *et al.* (1980b) considered using three-dimensional radar data to identify rainfall type, but reflectivity gradients may be large both in showers and widespread rain; and in thunderstorms they may be only a little greater than in some bright-band situations. The alternative is to use the temporal variability of the assessment factor as observed at a small number of calibration raingauges.

**THE OBJECTIVE DETERMINATION OF RAINFALL TYPE AND
THE CALIBRATION OF THE RADAR ESTIMATES OF
RAINFALL**

A simple parametric representation of the temporal variability of the assessment factor is required. However, Collier *et al.* (1983) found that the value of the standard deviation of the assessment factor over the previous 1 h does not discriminate amongst rainfall types sufficiently well to be of practical use. An alternative method related the rainfall type to the results of a harmonic analysis of the variations of running hourly mean assessment factors over periods of 1 h.

In general, raingauge data are available at 15-min intervals from at least four sites for the Hameldon Hill (NWRP) radar, and therefore a total of 16 assessment factors are calculated each hour. As two of the raingauge sites (Hollingworth Lake and Swineshaw, Figure 6.1) are only about 25 km apart, their assessment factors are too similar to be treated as independent. Therefore it was assumed that only three sites, providing 12 assessment factors per hour, are independent.

To eliminate from the analysis oscillations which might result from the separation of the raingauges—and relate more to raingauge siting differences than to variations of rainfall type—the observations available at each time, t, are meaned to give $A(t)$ where

$$A(t) = \tfrac{1}{3}\sum_1^3 A_i(t),$$

and $A_1(t)$ the assessment factor at site i.

The harmonic analysis is carried out over 1 h using four mean assessment factors at 15-min intervals, the series being given by

$$A(t) = \overline{A(t)} + \sum_{i=1}^{N/2} C_i \sin\left(\frac{360°}{P} it\right) + D_i \cos\left(\frac{360°}{P} it\right)$$

where $N = 4$ and $P = 60$ min.

This allows the evaluation of the coefficients of the two harmonics with periods of $\tfrac{1}{2}$ h and 1 h. However, in evaluating these coefficients it is assumed that the number of observations is twelve rather than four. This is an attempt to empirically scale the coefficients such that oscillations due to variations in rainfall types with periods less than $\tfrac{1}{2}$ h and greater than 1 h are indicated by reasonable values when compared with the total variance of the mean assessment factors (see below). These oscillations are likely to be manifest in the individual assessment factor variations, but may not be as evident in the variations of mean assessment factor over a period of 1 h; this scaling was therefore useful.

The calculation is carried out at time t for data at 15-min intervals over the previous hour. However, at $t-45$ min the data are assumed to be applicable to $t-55$ min, in order to reduce the influence of the assessment factors at the time farthest away from the current time.

The coefficients are evaluated as follows

$$C_1 = \frac{2}{N^*}\sum\left[A(t) \cdot \sin\left(\frac{360° t}{60\,\text{min}}\right)\right];$$

$$C_2 = 0 \text{ where } N^* = 12;\ t = 5, 30, 45, 60 \text{ min.}$$

$$D_1 = \frac{2}{N^*}\sum\left[A(t) \cdot \cos\left(\frac{360° t}{60\,\text{min}}\right)\right]; t = 5, 30, 45, 60 \text{ min}$$

$$D_2 = \frac{1}{N^*}\sum\left[A(t) \cdot \cos\left(\frac{360° t}{60\,\text{min}}\right)\right]; t = 5, 30, 45, 60 \text{ min.}$$

The percentage contribution that the two harmonics make to the total variance of the mean assessment factors is defined as V, where

$$V = \frac{\dfrac{Q_1^2}{2} + Q_2^2}{S_x^2}$$

$$Q_1^2 = C_1^2 + D_1^2; \; Q_2^2 = D_2^2$$

$\dfrac{S}{2}x$ = standard deviation derived from four values of $A(t)$ at 15-min intervals.

The value of V is taken as an objective measure of the variability. Normally the variance of the four mean assessment factors may be accounted for completely by the contribution of the two harmonics. However, the harmonics are evaluated assuming twelve observations, and therefore the contribution that the harmonics with periods $\frac{1}{2}$ h and 1 h make to the variance may take any value between 0 and 100 per cent. This procedure should be regarded as rough and ready, its success being judged by the discrimination between rainfall types it produces, rather than on its mathematical rigour. If the two harmonics, having periods of $\frac{1}{2}$ h and 1 h, account for most of the variability, then it is assumed that the rainfall type is frontal. On the other hand if the first two harmonics account for little of the variability, then it is assumed that variations are rapid and the rainfall type is showers. In order to discriminate bright-band and Welsh shadow conditions within the frontal category and from showers, the mean assessment factor over the hour is tested in conjunction with the value of V. Scatter diagrams of the mean assessment factor (A) as a function of V (multiplied by 100) for approximately 5 months of data (about 300 hours of data, September 1980–January 1981) have been prepared (Collier *et al.*, 1983). These scatter diagrams are the basis of the rainfall classification. There is a significant overlap between the showers, frontal rainfall and bright-band categories, but the probability of the different rainfall types can be assessed from the values of V and mean assessment factor. From these data the most likely rainfall type at locations over the area may be identified. The boundaries between types are produced objectively from probability values.

Frontal situations

Having recognized the rainfall type it becomes feasible to apply different calibration factors to radar measurements. Spurious assessment factors (perhaps derived from echoes arising from ground clutter produced in anomalous propagation conditions or from malfunction of the radar) are dealt with by applying limits outside which the values are not used (>3 and <0.3). These limits have also been applied to prevent large sudden changes in calibration factors which cause discontinuities in the appearance of the radar data. However, it was found necessary to relax these limits in the light of experience (see below). Calibration *domains* have been defined assuming westerly winds (Figure 6.7). The evidence for these domains has been presented by Hill *et al.* (1981) and Hill (1983). Generally, in frontal rainfall the necessary radar calibration is different on the windward and the lee slopes of hills. However, although frontal rainfall may be identified successfully, no way of differentiating objectively on-site between west wind cases and east wind cases has yet been specified. In the north-west of England few cases of significant rainfall with easterly winds occur (two cases in the 6-month period from September 1980 to February 1981). It may prove possible to separate west wind and east wind cases by comparing the assessment factors derived for raingauge sites on slopes exposed to west and east winds. Ways of using aneomometer data should also be investigated.

Ideally, one telemetry raingauge should be sited in each orographic domain. Until such time as this is possible, available gauges have to be used to estimate the assessment factor in other domains. If data from one or two gauges are missing the remaining values are weighted according to distance from the domain. This procedure is carried out for domains A and B in Figure 6.7. Outside the domains it was found that $Z = 200 \, R^{1.6}$ was an appropriate relationship to use and provided, on average, more accurate results than using a mean calibration factor, derived in real time from the raingauges. This may not be the situation for other radar sites in other geographical locations.

Frontal situations with Welsh shadow effect

Hill (1983) identified the effects produced by the interaction of strong south-westerly winds at low levels of the atmosphere with the mountains of Snowdonia in North Wales. In such a situation the domains are defined as in Figure 6.7, where it is recognized that the performance of the radar in measuring the surface rainfall over the southern Pennines will be quite different from that further north.

Frontal situations with bright band at close range

Over the past 10 years or so, work has been carried out to investigate techniques for removing the artificial enhancement produced in the radar estimate of surface rainfall when the radar beam intersects the level at which snow melts to form rain. At far ranges the radar bright band often occupies a small proportion of the beam width, and therefore the enhancement is quite small, the effect on the surface measurement accuracy not being too serious. However, within 100 km range of the radar site, bright band effects may cause severe deterioration in the accuracy of the estimates of surface rainfall (Harrold *et al.*, 1974; Central Water Planning Unit, 1977). Until a reliable technique (which may involve the use of telemetering raingauge data) is evolved, consideration must be given as to how best to use calibration raingauge data in bright-band conditions. The four raingauge sites to be used to calibrate Hameldon Hill radar measurements of surface rainfall are situated deliberately at different ranges from the radar. (As noted earlier, only three of these sites provide independent measurements for rainfall type identification; however, all four sites may be used to calibrate the radar data given the rainfall type.) The effect of the bright band is to enhance the radar measurements in either a full or broken annulus at a range from the radar site dependent on the radar beam elevation and the height of the melting level. Therefore, calibration factors derived for each raingauge site should be applied to the radar data in annuli around the radar site out to a maximum range of 50 km, as shown in Figure 6.7. Bright-band effects at further ranges from the radar are not dealt with. It is unwise to infer corrections in areas where no calibration raingauges exist in rainfall situations of great spatial variability.

This technique will not always be successful, indeed it could produce corrections *at a point* which lead to even worse answers than before the correction was made, for example, in cases when the bright-band height is changing rapidly (as a frontal zone moves over the area). However, the overall data accuracy should improve, as will the appearance of the distribution of surface precipitation, and this procedure may be regarded as a very simple form of bright-band correction procedure. More recently, experience has indicated that annular domains, whilst working well on occasions, can lead to dramatic modifications to the data. The accuracy discussed in the next section is applicable to annular domains, but it has now been decided to move, in bright-band situations, to the more conservative domains specified for showers.

Showers

Wilson and Brandes (1979) note that showers within a given location require different $R : Z$ relationships. Using only a few calibration raingauges it is not possible to produce correction factors for each shower. It would be unwise to define large domains around the existing calibration raingauges (Figure 6.7). In areas away from the small domains, a mean $R : Z$ relationship ($Z = 300\ R^{1.6}$), which has been found to be generally applicable for showers in north-west England, is used to calibrate the radar measurement.

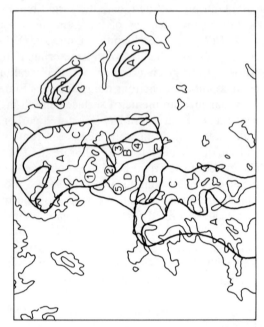

(i)   FRONTAL

Domain A   Use average calibration factors from gauges 1, 2, 3 & 4

Domain B   Use Z = 180 R 1.6

Domain D   Use calibration factors from gauge 5

Elsewhere   Use Z = 200 R 1.6

(ii)   WELSH SHADOW

Domain A   Use average calibration factors from gauges 1 & 2

Domain B   Use average calibration factors from gauges 3 & 4

Domain C   Use Z = 180 R 1.6

Domain D   Use calibration factors from gauge 5

Elsewhere   Use Z = 200 R 1.6

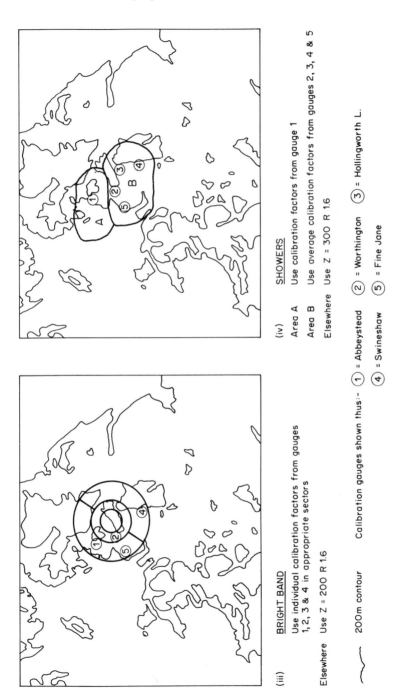

(iii)   BRIGHT BAND

Use individual calibration factors from gauges
1, 2, 3 & 4 in appropriate sectors

Elsewhere   Use Z = 200 R 1.6

(iv)   SHOWERS

Area A   Use calibration factors from gauge 1

Area B   Use average calibration factors from gauges 2, 3, 4 & 5

Elsewhere   Use Z = 300 R 1.6

⌒   200m contour   Calibration gauges shown thus:- ① = Abbeystead   ② = Worthington   ③ = Hollingworth L.

④ = Swineshaw   ⑤ = Fine Jane

Figure 6.7.   Domains used for different rainfall types and the corrections to the radar data made in each domain

THE ACCURACY OF THE RADAR ESTIMATES OF RAINFALL

It has been demonstrated that this domain calibration procedure produces rainfall estimates which are more accurate than those derived using uncalibrated radar data, and also more accurate than methods using a single mean calibration factor (Collier *et al.*, 1983). In order to assess the performance of this procedure, 12 months' data, up to January 1983, have been analysed. The results have been summarized in detail by Collier (1986a).

The true rainfall at a point, or over an area, is of course unknown. In the study reported here it has been assumed that the raingauge data are correct, unless errors are so large as to be obvious. Effort has been concentrated on examining hourly rainfall totals at NWWA telemetry raingauge sites (other than the calibration gauges) plus one or two additional Meteorological Office sites (see Figure 6.1).

Seasonal variations in accuracy

Figure 6.8 shows the variation of the monthly mean assessment factors over the 12-month period for four raingauge sites at different ranges from the radar. Three of these sites are within the area covered by the real-time calibration system, and Nottingham is included to illustrate the performance at greater range.

During the summer months, when deep convective rainfall was common, the radar estimated rainfall very well at all ranges. However, the average performance was not so good during the winter. The record for Nottingham shows the under-estimation which can occur at far ranges in winter due to the development of low-level rainfall below the radar beam, whilst the other graphs indicate a tendency to over-estimate at near ranges, probably as a result of uncorrected bright-band effects within the radar beam.

Errors in radar estimates of hourly rainfall

At any time of the year the assessment factors at individual rain gauge sites may vary considerably from the mean monthly values discussed in the previous section. When these variations are examined, the 'bright-band' effect is seen to be very significant. If all observations influenced by bright band are excluded, the average percentage difference between hourly uncalibrated radar estimates and rain gauge readings in conditions of widespread frontal rainfall is 60 per cent. Real-time calibration, as summarized in Figure 6.7, reduced this average difference to 45 per cent. In convective rainfall, without bright band, calibration improves the figures from 37 per cent to 21 per cent. When bright-band effects are present the average difference is 100 per cent and calibration reduced this to 75 per cent. These results compare

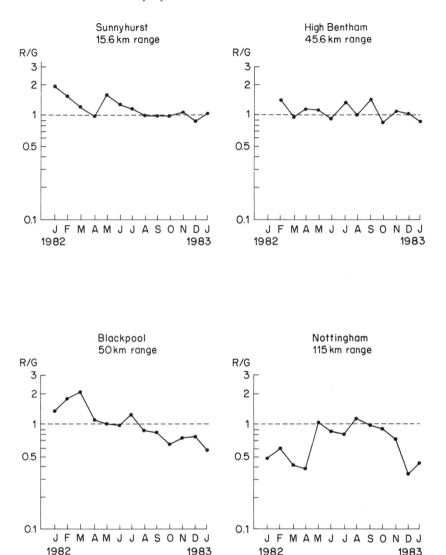

Figure 6.8.   Variation of monthly mean hourly assessment factors for four gauge
sites at different ranges from the radar

well with those achieved by off-line analysis in the Dee project (DWRP).

It is worth noting the fact that at the High Bentham raingauge site bright-band effects were present for more than 50 per cent of the time, and the development of methods for improving data quality when bright band is affecting the beam is therefore an important area for further work. It is likely that significant improvements in this area will only be attained using methods based on the vertical profile of radar reflectivity, and one such technique being developed by the UK Meteorological Office shows considerable promise (Smith, 1986). Figure 6.9 shows an example of the improvement possible using this technique, based upon a reflectivity height profile.

Comparison of hourly rainfall estimates from radar and telemetering raingauges

The NWWA telemetering raingauge network was designed to complement the Hameldon Hill radar installation, five of the gauges being sited so as to act as calibration gauges for the radar, and 13 others being located in areas where radar coverage was expected to be poor due to excessive range and/ or screening by intervening hills. The average density of the network is less than one gauge per 1000 km^2, and at least 50 per cent more gauges would have been included in the network if the radar had not been available.

The performance of the existing telemetering gauge network in measuring hourly point rainfall has been assessed by Collier (1986b). The bias and random errors which arise when using the raingauge data alone to estimate hourly rainfall at a point have been examined as a function of the distance of that point from the raingauge used to make the estimate. Similarly the errors in the radar estimates of hourly rainfall at a point were considered in relation to the distance of the point from the nearest calibration raingauge.

By comparing the distances within which, or greater than which, the gauges had lower random and bias errors than the radar estimates, it was possible to define where the telemetry gauge network is likely to outperform the calibrated radar. In general those areas are close to the gauges themselves, and at far ranges from the radar. The analysis is summarized in Figure 6.10, which shows the preferred methods for real-time estimation of hourly rainfall over the NWW area. To provide an equivalent database within 75 km of the radar site to that provided by the calibrated radar, an additional 40 or more telemetering gauges would be needed at a capital cost which would approach that of the radar system, if data processing costs were included. Such a dense gauge network would, however, not provide the benefits to short-period forecasting which the radar will bring by its detection capability at far ranges.

Operational problems associated with raingauge calibration

The real-time calibration procedure operated effectively for most of the time during the period covered by this assessment. In the previous section it was noted that, on average, the calibration procedure improves the accuracy of measurement of hourly rainfall. However, there has been a small number

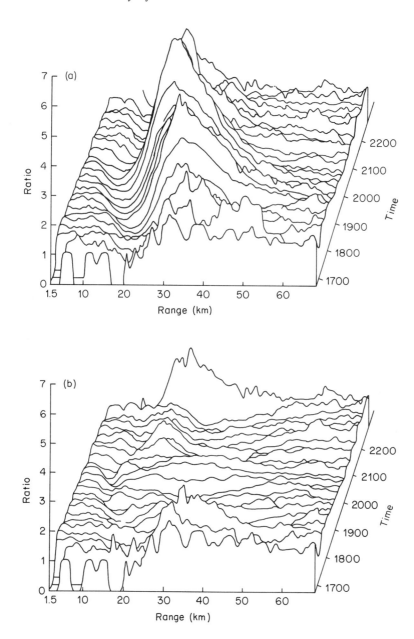

Figure 6.9. A time-history of ratio (lower divided by upper beam rainfall rates) versus range profiles (a) before and (b) after correction for 1645–2245 GMT, 9 December 1982, Clee Hill radar (Smith, 1986)

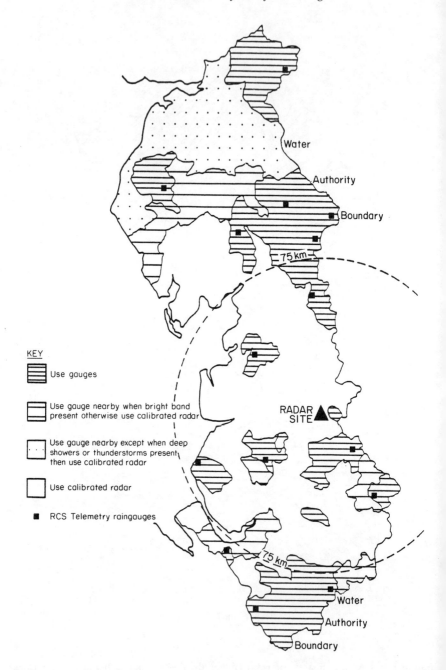

Figure 6.10.   Preferred method of rainfall estimation in the North West Water
Authority region

of occasions when the procedure has had an unexpected detrimental effect on the rainfall estimates. Equipment failures such as faulty rain gauge interrogations can have such an effect, but they have been very rare.

Operational problems with the calibration procedure have resulted from sudden changes in the perceived rainfall-type, sudden onset or cessation of calibration, and from the limits initially applied to the assessment factor in an attempt to reduce the visual impact on the TV display resulting from such events. Initially, only assessment factors within the range 0.3 to 3 were used for calibration so that, for example, in a bright-band situation where the uncalibrated radar might be over-estimating by a factor of 10, the calibrated information would still be over-estimated by a factor of more than 3.

The procedure was later refined to provide for more gradual introduction and cessation of calibration as rainfall passes over the calibration gauges, the assessment factor limits were widened to 0.2 and 10, and further work is in hand to stabilize the identification of rainfall type. It may be, however, that further statistical analysis of the A time series could provide more reliable estimates of precipitation type.

CONCLUSIONS

A raingauge calibration procedure has been successfully introduced into an unmanned fully automatic radar system. Calibrated radar data are provided in real time to both meteorological and hydrological users.

With the present configuration of calibration raingauges it is suggested that the calibrated radar estimates of point rainfall are better than estimates made using the NWW telemetry raingauges alone in frontal rainfall at distances up to between about 60 km from the gauges, depending upon the presence or absence of the bright band in the radar beam (Collier, 1986a). This means that useful measurements are possible to at least 75 km range from the radar site over the NWW area, but not over the Yorkshire Water Authority area where calibration is not yet applied. The calibrated estimates are more accurate than the raingauge estimates at all ranges in convective rainfall except very close to the raingauge sites (Collier, 1986b).

The number of raingauges required to calibrate a radar used within a quantitative hydrological forecasting system will depend on the area over which quantitative data are required, the nature of the terrain over which measurements must be made, and on the rainfall types which are most likely to produce floods. In the UK, both convective and frontal rainfall may result in flood conditions and the terrain is generally very variable.

It is suggested that ideally at least one calibration gauge should be placed in each calibration domain. However, for the Hameldon Hill radar, even though there are six domains within 75 km range of the radar, it had been found that five calibration gauges provide satisfactory data quality. If gauges were to be evenly distributed then it should be possible to produce reliable calibration in convective rainfall most of the time.

It has been shown that calibrated radar data provide more accurate estimates of hourly rainfall over a large area of the NWWA than those obtained using the present telemetry raingauge networks, particularly when bright-band effects are absent, and in convective rainfall. However, on occasions inappropriate calibrations can be produced, leading sometimes to large errors in the rainfall estimates which may have a significant detrimental effect on river hydrographs produced using the radar data as one input (Collier and Knowles; 1986). It is important that ways of exercising quality control, objectively, subjectively, or more likely by a combination of both, on the radar estimates of rainfall are investigated, so that random errors may be reduced further. Analysis of the relative importance of particular errors and their effects on river hydrographs will be needed to optimize these procedures. The FRONTIERS system (Sargent, Chapter 3 and Browning, Chapter 16 this volume; Browning, 1986) presently under test in the UK Meteorological Office, should enable assessment of the degree to which these quality control problems can be overcome.

REFERENCES

Batton, L. J. (1973). *Radar Observations of the Atmosphere*. University of Chicago Press, Chicago and London.

Brandes, E. A. (1975). Optimizing rainfall estimated with the aid of radar. *J. App. Met.*, 1339–45.

Browning, K. A. (1981). A total system approach to a weather radar network. *Proc. Hamburg Conf. of Nowcasting, WMO/ESA/IAMAP, ESA SP-165*, pp. 115–22.

Browning, K. A. (1986). Weather radar and FRONTIERS. *Weather*, **41**(1), 6–16.

Cain, D. E., and Smith, P. L. (1976). Operational adjustment of radar estimated rainfall with raingauge data: a statistical evaluation. Preprints, 17th Conf. Radar Met. Seattle, AMS, Boston, pp. 533–8.

Central Water Planning Unit (1977). *Dee Weather Radar and Real-time Hydrological Forecasting Project*. CWPU Reading, November, 172 pp.

Collier, C. G. (1986a). Accuracy of rainfall estimates by radar. (a) Part I: Calibration by telemetering raingauges. (b) Part II: Comparison with raingauge network. *J. Hydrol.*, **83**, 207–235.

Collier, C. G., and Knowles, J. M. (1986). Part III: Application for Short-term flood forecasting. *J. Hydrol.* **83**, 207–49.

Collier, C. G., Cole, J. A., and Robertson, R. B. (1980a). The North-West Weather Radar Project: the establishment of a weather radar system for hydrological forecasting. Hydrological Forecasting, *Proc. Oxford Symp.* April, IAHS Publ. No. 129, pp. 31–40.

Collier, C. G., Lovejoy, S., and Austin, G. L. (1980b). Analysis of bright-bands from 3-D radar data. Preprint Volume, 19th Conf. on Radar Met., Miami Beach, Florida, 15–18 April 1980, *Am. Met. Soc. Boston*, pp. 44–47.

Collier, C. G., Larke, P. R., and May, B. R. (1983). A weather radar correction procedure for real-time estimation of surface rainfall. *Quart. J. R. Met. Soc.,* **109**, 589–608.

Harrold, T. W., English, E. J., and Nicholass, C. A. (1974). The accuracy of radar derived rainfall measurements in hilly terrain. *Quart. J. R. Met. Soc.,* **100**, 331–50.

Hill, F. F. (1983). The use of average annual rainfall to derive estimates of orographic enhancement of frontal rain over England and Wales for different wind directions. *J. Climatol.,* **3**, 113–29.

Hill, F. F., Browning, K. A. and Bader, M. J. (1981). Radar and raingauge observations of orographic rain over South Wales. *Quart. J. R. Met. Soc.* **107**, 643–70.

Joss, J. K., Schran, J. C., Thams, J. C., and Waldvogel, A. (1970). On the quantitative determination of precipitation by radar. *Wissenschaftliche Mitfeilung,* No. 63, Eidgenossischen Konmission Zum Studium Hagelgilburg und der Hagelawhar, 33 pp.

Smith, C. S. (1986). The reduction of errors caused by bright-band in quantitative rainfall measurements made using radar. *J. Atm. Ocean. Tech.* (*in press*).

Wilson, J. W. (1970). Integration of radar and rainfall data for improved rainfall measurement. *J. App. Met.* **9**, 489–97.

Wilson, J. W., and Brandes, E. A. (1979). Radar measurement of rainfall—a summary. *Bull. Am. Met. Soc.,* **60**, 1048–58.

Weather Radar and Flood Forecasting
Edited by V.K. Collinge and C. Kirby
© 1987 John Wiley & Sons Ltd.

CHAPTER 7

Weather Radar and a Regional Forecasting Service

F. DALTON

The Manchester Weather Centre is the main forecasting office for the north-west region of England, located in Stockport within the Greater Manchester area and with responsibilities towards civil aviation as well as public services. The duty staff are on continuous watch throughout the 24 hours, 7 days per week and have access to a Jasmin type radar display. Manchester lies on a plain facing to the west and north-west towards the Irish Sea (Figure 7.1), and protected to some extent through an arc extending from 330° through north to 150° by hills forming part of the Pennine chain, which lies 20–25 miles (30–40 km) away. To the south the ground gently rises across the Cheshire Plain towards the Midlands, while to the south-west some 40–50 miles (65–70 km) distant lie the Welsh mountains which also play a significant part in the variation of the weather patterns across this area.

These regional variations in the weather are well documented in reference manuals and in papers written by Weather Centre staff. As a consequence forecasters new to the area are quickly able to familiarize themselves with local peculiarities that can result from the orographic influences on the effects of rain, low cloud, or wind directions. This of course is true for most stations throughout the UK. In setting about the task of studying the weather pattern on a particular day the forecaster refers to his synoptic charts. These charts display in a plotted symbol format the weather reports at all the weather stations over the area taken at the same hour of the day.

Figure 7.1 shows the distribution of synoptic reporting stations over the north-west region of England from which the forecaster is able to gain some intelligence as to the distribution of significant weather around his area. It is easy to see that this sparse network of reports leaves large areas over

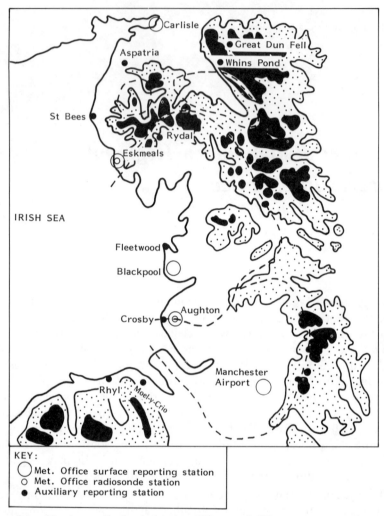

Figure 7.1. Northwest England showing 200 m and 500 m contours and synoptic
stations

which the forecaster will have no knowledge. It is a wise policy always to
ask an enquirer, telephoning for information about the weather prospects in
his area, for a brief report on the weather conditions at the time of his call.
The weather often holds a surprise in the variations that occur over hilly
districts. In addition, the forecaster will hope to glean some insight into the
vertical structure of the atmosphere by studying the radio-sonde information
which gives a vertical profile of the change of temperature, humidity and
wind against pressure over the area around the reporting station. The nearest

radio-sonde station to Manchester is found at Aughton which lies near the coast to the west of Manchester and approximately 8 miles (13 km) north of Liverpool. The radio-sonde stations release the balloon-borne package of instruments on a routine 12-hourly schedule based on midnight and midday.

Because of the problems created by a sparse network of synoptic observations on a global scale, a broad programme of research has been undertaken during the past 20 years which has resulted in rapid development of remote sensing instrumentation. Various forms are now available to the forecaster at the bench, and include such items as weather radar, satellite cloud pictures, and automatic weather stations both on land and on the sea.

The forecasting staff at the Manchester Weather Centre receive information from each of these sources and they are of great benefit. Automatic weather station reports are collected from six stations in the region, a comprehensive schedule of satellite pictures is received throughout the day from the Meteorological Office at Bracknell, and in addition the forecaster is able to refer constantly to the radar display. The Jasmin type display was installed in November 1980. Initially it gave the distribution of echoes around the Hameldon Hill radar site only, but this has now been extended and the display incorporates the composite picture on the national grid (see Figure 1.2).

The forecasters based at Manchester supply information to the North West Water Authority in the form of expected events of significant rainfall over the catchments in the region, a service which extends back over 20 years before the use of radar. Within this period the forecasters have devised techniques and knowledge of how to judge which synoptic weather patterns will likely result in a significant fall of rain across parts of the region. The main problem is centred on the orographic enhancement of the rainfall by the high ground. In this context the important parameters have been found to be the available amount of moisture in the prevailing air mass below 1200 ft (360 m) and the strength and direction of the wind flow pattern in this lower part of the troposphere (Figure 7.2). It is evident from the distribution of catchment areas in Figure 7.3 that moist airstreams approaching from the west or south-west offer the main threat of prolonged rainfall across the region.

Most events of prolonged significant rainfall occur in association with Atlantic depressions moving close to the north-west of the British Isles. This means that when such a pattern occurs on the Atlantic weather chart the duty forecaster must be aware of the possibilities, and he has then to assess how this particular pattern will evolve as it crosses into western parts of the UK. Initially he must decide whether or not the characteristics of the air mass will fulfil the criteria for a rainfall warning. In these decisions the guidance available to him is not the radar display but evidence of the characteristics of the air mass gleaned from synoptic reports (ship obser-

Weather radar and flood forecasting

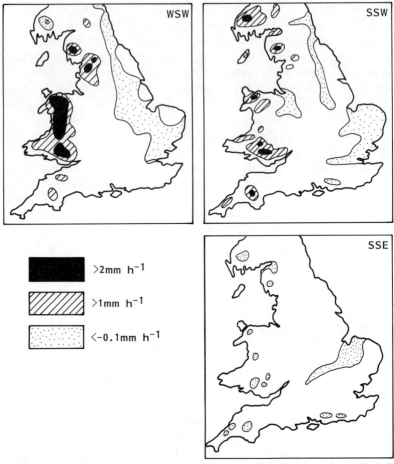

Figure 7.2.   Orographically induced distribution of rainfall with various wind directions

vations), or any radio-sonde information which might be available within the airstream, but more significantly evidence compiled from the structure of the cloud patterns in the developing weather system. In addition to these various forms of observation the forecaster can also refer to the computer-derived forecast fields, showing predictions of the pressure patterns and expected rainfall along with an indirect measure of potential moisture in the airflow over the region. From these various assessments the duty forecaster will make a judgement as to whether or not to issue a warning to the water authorities. In turn the radar will become a vital and useful tool in the constant monitoring of the situation as it progresses over and across the region. From his knowledge of the synoptic weather pattern and his obser-

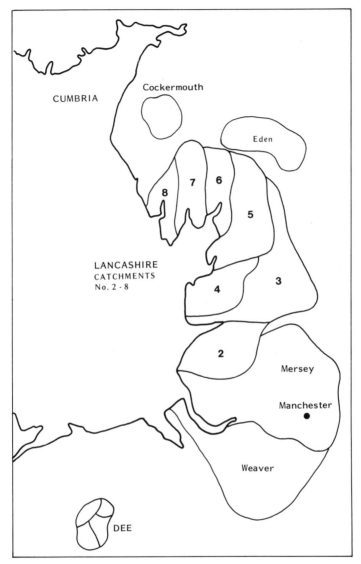

Figure 7.3. North-west England catchment areas

vations of the distribution and intensity of the radar echoes as the situation develops into the region, the forecaster can update his warning and liaise with the appropriate water authority. Thus the radar has become a vital source of identifying significant rainfall occurrences in a particular district within the time scale of 2–3 h ahead. As an offshoot to this the information, which is of course very useful to the water authority staff responsible for flood forecasting, also provides the forecaster with additional insight into the development of rainfall areas across his own region, and therefore adds to his fund of knowledge of the local area.

A more direct and appropriate use of the radar in identifying the likely event of significant rainfall is in the situations leading to local heavy showers or thunderstorms. Heavy summer-type showers and thunderstorms are a feature of a slack synoptic pattern, and can be identified as developing frequently over the west Midlands areas and Welsh borders. High afternoon temperatures occur over these sheltered inland districts which may result, if the atmospheric conditions are favourable, in the development of large cumulus or cumulonimbus clouds. The development and growth of these clouds can be enhanced over the area by local convergence within the wind field. Such a convergence may be as the result of an afternoon sea breeze moving inland across the South Lancashire plain and over Cheshire running up against the circulation caused by a heat low over the hot inland areas of the Midlands. Similar convergence of winds can also result from katabatic winds developing overnight as cold air drains out of the Welsh mountains; this could help to perpetuate storm clouds which have developed the previous (late) afternoon or evening. Considerations of this sort illustrate how storm clouds can affect the region around Manchester due to the geographical features of the area. The possibility of these events can be identified but the exact location is not so easy to predict, nor is the intensity of the rainfall. It is in these situations that the radar has proved itself to be an essential source of information.

An example of such a case occurred on the evening of 5 August 1981. An unstable air mass covered England on this day ahead of a developing upper trough from the Atlantic, moving slowly into the western approaches of the UK (Figure 7.4). There had been a day or two of increasingly humid weather over France, and as a result of the convergence ahead of the upper trough and the moist air advecting northwards from France, storm clouds developed over a wide area of the south of England moving north-north-eastwards. The forecasters at Manchester were well aware of the developing situation and were keeping a watchful eye on thunderstorms reported and observed over the Midlands during the afternoon. However no rainfall warning was issued to the water authorities that afternoon. At 1950 GMT a line of rainfall cells was observed on the radar (Figure 7.5) extending from Oswestry (on the Welsh border) in the west, to Spalding (near the Wash) in the east, with several cells indicating 32 mm rainfall equivalents and a few of equivalent 16 mm rainfall. By this time the duty forecaster at Manchester had already issued a thunderstorm warning for the region but now, on the evidence of the radar, an alert was given at 1955 GMT to the duty hydrologist of the Mersey and Weaver Division (see Figure 7.3). The developing weather

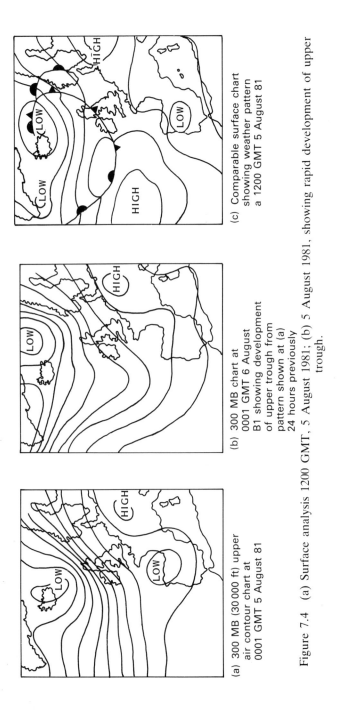

(a) 300 MB (30 000 ft) upper air contour chart at 0001 GMT 5 August 81

(b) 300 MB chart at 0001 GMT 6 August B1 showing development of upper trough from pattern shown at (a) 24 hours previously

(c) Comparable surface chart showing weather pattern a 1200 GMT 5 August 81

Figure 7.4 (a) Surface analysis 1200 GMT, 5 August 1981; (b) 5 August 1981, showing rapid development of upper trough.

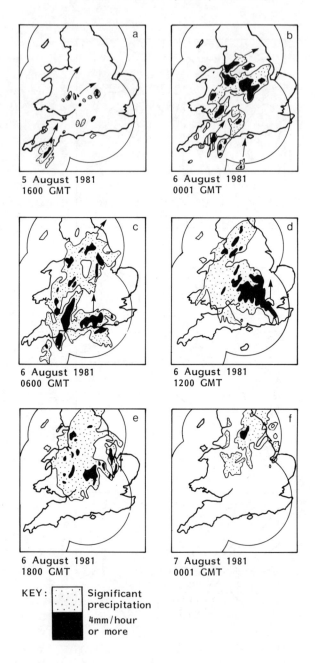

Figure 7.5. Storms of 5/6 August 1981, Jasmin display.

pattern was constantly monitored by the staff on duty in the forecast room and at 2040 GMT warnings were given to the air traffic control authorities at Manchester Airport concerning the impending threat of storms across the area with likely effects to aircraft operators. The rain, ahead of the main thunderstorm cells, started over the airport around 2210 GMT and by 2347 GMT this had developed into a vast spread of rain some 60 or 100 km wide extending from the Welsh border across to Lincolnshire. At 0100 GMT the duty forecaster once again contacted the duty hydrologist to report an area of very heavy rain with thunderstorms centred just west of the airport towards Warrington, covering a radius of around 20 km. This area of heavy rain was predicted to continue moving slowly north-east with rainfall amounts expected to exceed 30 mm in places during the next few hours. This warning was proved to be fully justified, for at Manchester Airport a total rainfall count of 87.6 mm (3.5 in) was recorded between 2200 GMT and 0300 GMT on the morning of the 6th. The return period for an event of this amount of rain is in the order of 200 years.

Another more recent occasion on which the radar was of invaluable help to the duty forecaster occurred during the evening of the 14 May 1985. On this occasion an area of low pressure covered the southern North Sea with its circulation of easterly winds across northern England. The air mass was unstable and outbreaks of thundery rain were observed on the synoptic chart reports and were clearly evident from the intensity of the radar echoes. The gradient across the NWWA catchment areas was east-north-easterly, of the order of 15 m s^{-1} (approximately 30 knots); under the normal criteria for heavy rainfall over the water authority's areas this direction of wind flow is not considered significant for heavy rain. However, during the evening heavy rain developing over the Lancashire catchments was indicated by the radar, and the duty forecaster alerted the duty hydrologist of the Lancashire and South Cumbrian Division (Figure 7.3). This was at 1745 GMT, with the warning given for the period up until 2300 GMT. The rainfall totals recorded on this occasion fully justified the warning given, with the evidence from the radar being the only indicator available to the forecaster since the normal predictors do not cover this form of development of the weather pattern. The synoptic network over the Pennines as shown in Figure 7.1 is certainly not dense enough to give a clear indication of rainfall spreading in this way across the area.

It is not only in predicting rainfall events that the radar is a vital source of information to the forecaster in the North-West region. One particular weather pattern that is a common feature, and of some significance, is the occurrence of a cool polar maritime north-westerly wind flow. Under favourable atmospheric conditions lines of shower clouds develop along the streamlines in the flow pattern over the Irish Sea and run inland across the coast. Often these developments can be detected on the satellite pictures

(Figure 16.10) but these cannot give sufficient detail as to the path of the shower clouds as they move inland over the Lancashire coast. Narrow bands of almost persistent showers are experienced under these circumstances, and whilst the rainfall may not be significant enough for a warning to the river authorities, for aviation purposes the regime is important because of a lowering of visibility and cloudbase that often occurs under the path of the cloud formations. The aviation forecaster at Manchester can make full use of the radar display (Figure 16.5) to locate the trajectories of the shower clouds, particularly the movements—speed and directions—of the cells, to give him more accurate timings of likely visibility and cloud base fluctuations over airfields in the region under his supervision. This is especially true when giving a TREND message on the end of the weather report for the airfield. This TREND message indicates likely fluctuations in the 2 h period following the observation.

Another particular feature of the north-west region is the marked lowering of cloud behind the movement of a cold front over the area. In normal circumstances the veer of wind behind a cold front is often accompanied by improving weather as cooler polar air follows, but this is frequently not the case in the Manchester area due to the surrounding geographical features. In the south-westerly or southerly wind flow ahead of a cold front moving into the area from the west, the Welsh mountains and the land track afforded from that sector frequently result in better visibilities and cloud base reports than observed elsewhere in the moist warm sector feed of air. However, as the cold front progresses further east and moves across the Lancashire plain towards Manchester and beyond, the resulting veer of wind draws air across the region from off the Irish Sea. This unsheltered flow brings the moist sea air with it, and progressing over what may be cooler land—particularly in the case of a winter situation—leads to an unexpectedly prolonged period of low cloud and reduced visibility behind the front instead of the normally anticipated improvement. The forecaster using the radar can be fully aware of the progression of the rain area associated with the cold front, and therefore can anticipate in advance a more accurate assessment of the change of circumstances that the front will bring.

Cold fronts can be very varied in the type of weather associated with them, depending upon the structure of the tropospheric wind circulations existing at the time. The forecasters may refer to these loosely as 'ANA' type or 'KATA' type fronts. Usually an active front marked by strong vertical upward movements is termed an 'ANA' front and conversely, when the depth of cloud is limited by weaker upward motion or reversed flow resulting in downward vertical motion, the front is called a 'KATA' front. An 'ANA' cold front may be accompanied by very active embedded cells of unstable cloud, perhaps with thunderstorms, and this can lead to the front being marked by a squall line, or marked by an area of significant vertical wind shear close to the ground likely to affect aircraft landing or taking off. Under these circumstances it is imperative that the forecaster gives a positive indication to the airfield authorities of when and for how long the difficult and dangerous flying conditions are likely to affect the local area. The radar provides the information on the movement and development of such squall lines, allowing the forecaster to give accurate advice to the airfield authorities.

Cold fronts are also important features to follow because of the change to cooler weather that follows. During the winter months an important aspect of the forecaster's duties is to advise local authorities when circumstances may lead to a sudden fall of temperature overnight resulting in road temperatures dropping below freezing point. A cold front moving over the region during the evening can leave roads wet, and then be followed by freezing conditions, giving icy roads for the early-morning rush hour. The timing of the ending of the rain and ensuing fall of temperature is important to give correct advice to the borough engineers over when to 'salt' the road surface to alleviate the problem of ice formation. Once again the radar is a vital source of information for accurate advice.

There are many similar aspects of public service work where the benefits of the radar are indisputable. Among these are the forecasting the onset of rain for sporting events, farming enquiries, building work or live broadcast weather advice to the community as a whole. The forecasters in Manchester are very involved in all aspects of this type of work, and are therefore extremely grateful for the additional help and advice available to them through data from the radar network.

Weather Radar and Flood Forecasting
Edited by V.K. Collinge and C. Kirby
© 1987 John Wiley & Sons Ltd.

CHAPTER 8

An Operational Flood Warning System

G. A. NOONAN

INTRODUCTION

When the North-West Water Authority (NWWA) was formed in 1974 flood warning schemes were inherited from former river authorities (Figure 8.1). These schemes used rain and river level gauges which were interrogable by telephone and which gave data for use in simple—usually graphical—flood forecasting models. The river authorities had proposals to improve their flood warning facilities but these were withdrawn in favour of a regional flood warning scheme using an enhancement of the proposed radio and communications computer facilities that were to be set up for monitoring and controlling the Lancashire Conjunctive Use Water Supply Scheme. Combining flood warning with the Lancashire Conjunctive Use Scheme's communication and computer facilities had clear cost advantages and a regional communications scheme (RCS), which also included inter-divisional communications links, was implemented in 1979. There was sufficient capacity in the computer facilities to operate, in real time, relatively simple flood forecasting models using both routing and rainfall-runoff relationships. In 1980 the first unmanned weather radar station to be commissioned in the UK was linked into the communications system. This gave the ability to use in models real-time areal rainfall as observed and measured by radar. A further computer was added at a later date, which gave the necessary capacity for off-line development of flood forecasting models and for the use in real time of more complex models than could be used with the original RCS computers.

This chapter describes the RCS system as it is used in flood warning and the preparation and dissemination of flood warnings through the police. The

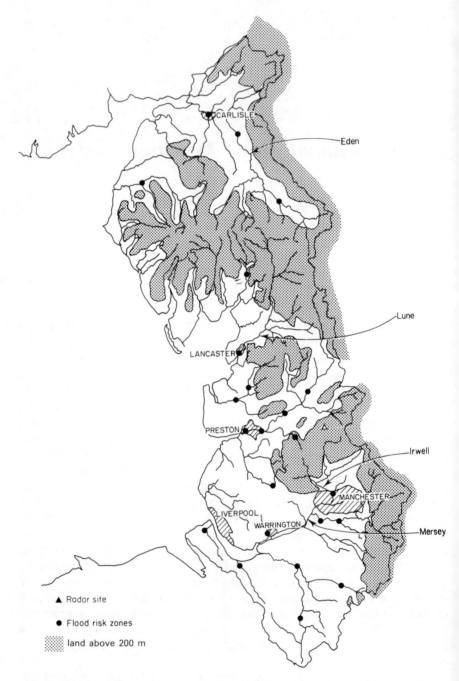

Figure 8.1.   North West Water—flood risk zones.

types of flood forecasting models used are referred to; these are discussed in more detail in Chapter 10.

THE FLOOD WARNING SYSTEM

Communications

The communications system consists of a microwave radio trunk route linking Carlisle and Warrington via the radar site and a number of telemetry scanning stations. Other scanning sites are connected via fixed UHF links. The system has channels for the monitoring and control of the Lancashire Conjunctive Use Scheme, inter-office telephone and computer links, mobile radio, and for hydrometric and radar data. The main computer system is situated at Franklaw and, for flood warning purposes, collects river levels, rain gauge readings and radar-derived subcatchment rainfall estimates every 15 min. The communication links allow a more recently installed computer at Warrington to be continuously updated with flood warning information. The outstations currently connected to the system include eighteen rain gauge sites, 49 river level stations and four tidal stations. These and the communication system are shown in Figure 8.2.

Computer system

Dual PDP 11/34 computers at Franklaw interrogate flood warning stations at 15-min intervals. All information is stored on both computers so that the standby machine can take over automatically in the event of a failure of the duty machine. In general, data are stored for 48 h. The computer system also provides facilities for simple arithmetic manipulation of the data on completion of each scan, and these facilities are used to produce forecasts of river levels at significant flood risk areas. The forecasts are produced automatically every 15 min and stored alongside the observed data. River levels, flows and forecasts, rainfall amounts and intensities can all be displayed in graphical or tabular form on terminals at the authority's office in Warrington. Printed reports can also be obtained over the public telephone network using a portable terminal fitted with an acoustic coupler, enabling a duty officer to obtain all essential flood warning information at home.

The additional computer, a further PDP 11/34, sited at Warrington, is continuously updated with the 15-min data collected and processed by the Franklaw computer. This computer gives the following facilities:

1. Additional 'phone-in' points for acoustic coupled computer terminals.
2. The ability to run, in real time, more complex models than the Franklaw computers.

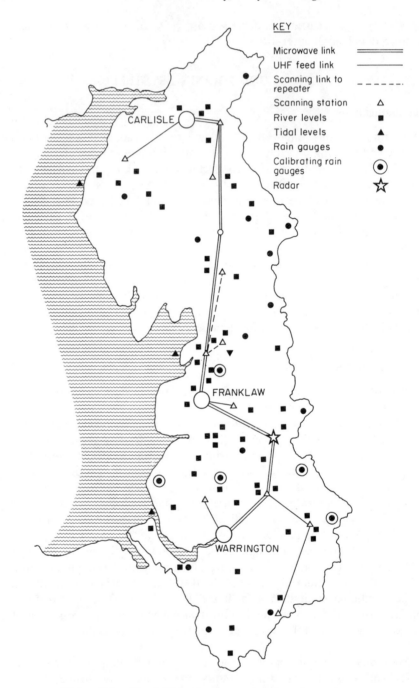

Figure 8.2. North-West Water communications system

3. The capacity to carry out off-line model development work.
4. Acceptance of data from solid state loggers.
5. Communications links to the authority's mainframe computer (ICL 2900 series) where hydrometric data are archived.

If required automatic dial-out facilities can readily be added to this computer should it be necessary to automatically interrogate outstations through the public telephone network.

Outstations

River level

Most of the river level outstations used for flood warning are conventional gauging stations where river level is reproduced in a well. The water level in the well is measured for flood warning purposes by floats coupled to encoders, or punched-tape recorders (PTRs) with encoders taken off their drive shafts, each producing binary coded digital signals which are transmitted by radio to the Franklaw computer. Some stations, generally used solely for flood warning, do not have a well, and river level is measured directly by transducer.

Raingauges

Tipping bucket raingauges are used, usually with a tipping volume equivalent to 0.2 mm. These data are converted to BCD for transmission. Two raingauges are used at each site. Comparisons between their outputs are made by the Franklaw computer and appropriate alarms are raised if the readings vary by more than a predetermined amount.

Weather radar

The radar, sited at Hameldon Hill near to Burnley, is a Plessey Type 45C, operating at C band (5.6 cm wavelength) and is intended to be capable of measuring peak rainfall rates over intensities of 0.1 mm h^{-1} to 100 mm h^{-1} at ranges of at least 75 km. Although basically a standard equipment it has been modified to operate in the remote environment of Hameldon Hill. A full description has been given by Hill and Robertson in Chapter 5.

When translating the radar signals to rainfall an automatic calibration process is carried out for varying types of rainfall conditions by reference, in real time, to data in the Franklaw computer from six of the interrogable raingauges.

The output from the radar is in two forms:

1. A VDU picture which shows the location of rainfall on a map of the northern part of England with colour coding to indicate rainfall intensity.
2. An areal measure of rainfall over some 100 subcatchment areas covering the North-West Water Authority area.

Both forms of data are updated every 15 min, the former being displayed in the Warrington Flood Forecast room (also at Yorkshire Water Authority and the Meteorological Office, Manchester), the latter being transmitted to the Franklaw computer for use in flood forecasting models.

Flood forecasting models

It is possible with the computer system, now that it has been enhanced by further computer capacity at Warrington, to operate a broad range of models from relatively simple routing and rainfall-runoff models to complex systems using a full spectrum of radar, raingauge and river level data with the provision for inserting forecast rainfall values.

Factors which have been taken into consideration when choosing a model for operational use are:

1. the accuracy of the forecast;
2. the time available between the preparation of the forecast and flooding taking place (lead time);
3. the manpower and computer resources required to operate the model under operational conditions.

A further point is that there appears to be little benefit in issuing public warnings more than about 4 h before flooding is likely to take place. If good communications are created to disseminate warnings quickly there is little point in increasing the complexity of a model to give lead times much in excess of 4 h—especially if a preliminary standby (not issued to the public) can be given to the police and local authorities so that they are ready to react quickly in the event of a specific flood alarm.

A description of the choice of suitable models and operation experience with them is given by Knowles in Chapter 10. At the time of writing (1986) current procedures use automatic, continuously operating models but without automatic real-time correction. However, the disadvantage is largely over-come by plotting river level forecasts for a flood zone, and using these as a basis for extrapolating observed river levels. This is discussed below under 'Preparation of forecasts'.

Preparation of flood warning schedules

A good deal of time and effort goes into the development of flood forecasting models, the development of field instruments and the communication of data from these instruments, in real time, into integrated flood forecasting systems. It is equally important to develop schedules of action to be taken when various river level stages in a potential flooding situation are predicted.

The responsibility for issuing flood warnings rests with the police, and it is their duty to co-ordinate any relief work that is necessary. In police forces it is said that because of shift work, weekend work and holidays, it is necessary to have about five officers to fill any one post. Further, officers tend to move about relatively frequently to other posts at different locations. The chances, therefore, of any one police officer gaining much experience of action to be taken in any particular area in the event of a flood are extremely remote. It follows that action schedules should be simple, explicit, unambiguous and based on the assumption that no-one who has to carry out the procedure has any experience in flood events. Equally, the 'standby' and 'alarm' messages to police which initiate the flood warning procedures and trigger progressive stages of action during a flood should be simple standard messages.

It is important to give the police adequate preliminary warnings that flooding might take place, therefore 'standby' messages are issued based on forecasts of river levels equivalent to about a two to three times/year frequency, even though flooding is known not to occur more frequently than say once in 20 years. Such early warnings enable the police to give similar early warnings to emergency services and local authorities without, at that stage, alerting the public. Issuing standby messages in this way has the advantage that communication lines between the water authorities, police, emergency services and local authorities are tested on a regular basis. Police do not consider this relatively frequent warning 'standby' to be 'crying wolf', but take it as a very necessary training action in communications so that when a 'real' event arises, messages are passed without confusion and delay. A typical schedule is shown in Figure 8.3. The police prepare their own more detailed warning schedules and action lists from the authority schedules. The maps accompanying the schedules show areas likely to flood at different levels of the river. Problems in estimating what a river will do when it comes out of its 'normal' channel make the theoretical preparation of the maps very difficult. If records are available of the extent of previous floods, particularly if the floods were of different magnitude, and if good hydrometric data has been retained from a river level/flow station in or near the flood risk zone—and it is of utmost importance that such an outstation should be available, not just for record purposes but for correcting forecasting models in real time—then the task of preparing the maps is relatively simple,

NWWA-RIVERS DIVISION	S.E. AREA	POLICE
FLOODWARNING SCHEDULE FOR RIVER IRWELL		SCHEDULE
FLOOD AREAS SALFORD		**GM 2**
		(includes 1 map)

AREA DUTY OFFICER MAY BE CONTACTED DURING OFFICE HOURS ON 061 973 2237 , NIGHTS AND WEEKENDS ASK FOR GREATER MANCHESTER DUTY OFFICER'S TEL. NO. FROM AUDENSHAW CONTROL, TEL. NO. 061 370 2645, 061 236 5829 OR 061 370 7511, OR FAILING THIS CONTACT ANY ONE OF:–

J. Blair	061 973 4896	J.R. Crellin	0606 41461
D.J. Wilkes	0925-75 3381	T.N. Linford	0925 65820
H.T. Davidson	0925 67950		

FLOOD WARNINGS WILL BE ISSUED TO GREATER MANCHESTER COUNTY POLICE TEL. NO. 061-236-8093/4 Telex 667897 or 667898 Local contact (No warning sent direct) – F Division – Salford 061 736 5877 (Extn. 232)

TYPICALWARNING:–	POLICE ACTION TO BE TAKEN:–
Flood Warning G.M.2 IRWELL STANDBY	This warning at least 3½ hours before flooding might take place. Advise Local Authorities as appropriate. Do NOT issue warnings to the public at this stage. Await further messages from N.W.W.A.
The warnings of flooding will be given about 2½ hours before the event Flood Warning G M 2 IRWELL ALARM (Level) A (likely to be reached at) . . . . hrs	Warn NOW Areas A on maps as Golf Course and Kersal Way likely to flood when level A reached.
. . . . .Do . . B . . . . . . hrs	Warn NOW Areas B on map as flooding of Lower Broughton area likely when level B is reached.
. . . . .Do . . C . . . . . . hrs	Warn NOW Areas C on map as flooding of streets and properties in Lower Kersal Areas likely when Level C is reached.
. . . . .Do . . D . . . . . . hrs	Close Cromwell Bridge (Point D on map) when level D is reached.
. . . . .Do . . E . . . . . . hrs	Warn NOW Area E on map as flooding of streets likely in Charlestown area when level E is reached.
. . . . .Do . . F . . . . . . hrs	Warn NOW all additional properties within the dotted area on the map. This area will flood when level F is reached.
Stand down	Flood peak over or danger of flooding past.

Figure 8.3. Typical flood warning schedule

although adjustments will have to be made to take into consideration alleviation works which are often carried out after flooding incidents. Even if recorded data of previous events is poor, it can often be used to supplement theoretical assessments of areas likely to flood. It follows that when a flood event does occur it is more important to record not only the extent of flooding but, by interviewing residents, shopkeepers, industrialists, etc., to obtain times when flooding started. By comparing these times with the recorded river level/flow occurring at the same time, it is relatively easy to determine the extent of flooding for various river stages. A graph for use by the duty officer is prepared for each flood risk zone and used in conjunction with the flood warning schedules.

Flood forecast room

Under normal conditions monitoring of rainfall and river levels is carried out by duty officers either in the flood forecast room at the Warrington office or from home, using portable computer terminals. When there is any possibility of tidal or inland flooding the flood forecasting room is continuously manned. The facilities available in the forecast room are:

1. Two-colour VDUs, which may be used to plot graphs and diagrams or display tables or rainfall, river levels and forecast river levels for all major flood risk zones. The VDUs may be coupled to either the Franklaw or the Warrington computers.
2. A further black-and-white VDU coupled to the Warrington computer for use with more complex interactive models and for development work.
3. A telefacsimile machine operating through the switched telephone network to receive surge forecasts from the storm tide warning service at Bracknell. This machine is also used for passing data to the authority's operations areas.
4. A graph plotter to copy any VDU picture.
5. A line printer which can give full details in tabular form of rainfall, river levels and river level forecasts. Rudimentary plots of data are also available on this line printer. Anything that can be produced on the line printer can also be obtained on the portable terminals, which may be connected into the computers by the switched telephone network.
6. A VDU giving the radar display together with a memory giving instantaneous replay of the last nine pictures (i.e. going back through the last 2 h).
7. A tape recorder to hold approximately 15 h of radar data as displayed on the VDU. This may be played back over the VDU but takes about 40 s for each picture to generate.
8. A satellite receiver picking up Meteosat data which gives near real-time

visual and infra-red pictures of cloud cover. These pictures enable duty officers to supplement forecast information supplied by the Meteorological Office.

Alerting of standby duty officers

Standby duty officers are alerted, in one of three ways:

1. The duty forecaster at Manchester Airport Meteorological Office issues heavy rainfall and gale warnings (for use in tide forecasts) by telephone (Chapter 7). To assist in forecasting he has available a radar display which shows either the picture from Hameldon Hill or the composite picture from Hameldon Hill, Clee Hill, Upavon and Camborne.
2. The computer at Franklaw continuously monitors river levels, weather radar, raingauges and forecasts from both routing- and rainfall-based models. Alarms are raised should any appropriate forecast exceed the NWWA standby level, which is usually about a five times/year ocurrence. The computer automatically raises the alarm at the Warrington office. If the alarm is not acknowledged (i.e. at night or weekend) the alarm is repeated at a continuously manned control room in a water supply station whose staff then advise the flood warning duty officer by telephone.
3. The duty officer at the Storm Tide Warning Service at Bracknell advises the Rivers Division duty officer if abnormal tidal surges are forecast.

Preparation of forecasts

On receipt of an alert, duty officers start to monitor the situation either from home or from the office as appropriate. Data from the control computer are available on request in report form for flood-risk zones. A typical report is shown in Figure 8.4. The reports list basic fixed parameters—NWWA standby level, police standby level and initial flood levels—together with current river level and forecasts generated by various models.

The current and previous river levels are plotted on the prepared graph (Figure 8.5) which also shows NWWA standby, police standby and various flood levels. Outputs from the various forecast models are also plotted. The trends of the model forecasts—which may well be high or low depending on antecedent catchment conditions, etc., are used as a basis for extrapolating the observed levels to give the best estimate of future river levels. Where possible, forecasts up to 4 h ahead are attempted. There is little benefit to be gained by forecasting river levels for more than 4 h ahead and, because of model limitations, periods rather shorter than this have to be accepted

REPORT "CEN"   29-JAN-85 15:26:54 ——|— Current Date and Time

FORECAST RIVER LEVELS – CENTRAL REGION

	NWWA STAND-BY	FLOOD LEVEL "A"	Past values	Current values	+1HR	Forecast values		
						+2HR	+3HR	+4HR
GARSTANG PS OBS.	1.70	2.60						
-F/C FROM ABBEY"D			2.20	2.22	2.19	1.96		
-F/C FROM RADAR			2.29	2.47				
ST. MICHAELS OBS.	3.75	4.50	3.07	3.58				
-F/C FROM ABBEY "D			1.91	2.98	3.97	3.71	3.09	
-F/C FROM RADAR			2.32	2.97	4.24	4.76	4.62	4.53
WIGAN OBS.	0.90	1.20	.711	.780				
-F/C FROM RADAR			.605	.627	.596	.575		
DARWEN OBS.	-.--	-.--	.953	.914				
-F/C FROM RADAR			2.16	1.87	1.67			
CLITHEROE "OBS".	-.--	-.--						
-F/C FROM RADAR								
RIBCHESTER "OBS":	2.50	3.60						
—— FROM JUMBLES			1.73	2.07	2.22	1.92	1.88	
-F/C FROM RADAR			1.35	1.80	1.94			
WALTON KC. "OBS".	7.50	8.50	5.83	6.23	6.49	6.97	7.09	6.84
—— FROM SAML "BY			6.20	6.39	6.67	6.73	6.86	
-F/C FROM JUMBLES			5.74	5.87	6.34			
-F/C FROM RADAR								

LAST SCAN AT: 1345   JMK 18-3-83
Status of data —|— ("I" –INVALID   "S" –SUSPECT)

F/C – Forecast Models

END OF REPORT

Figure 8.4. Example of forecast printout

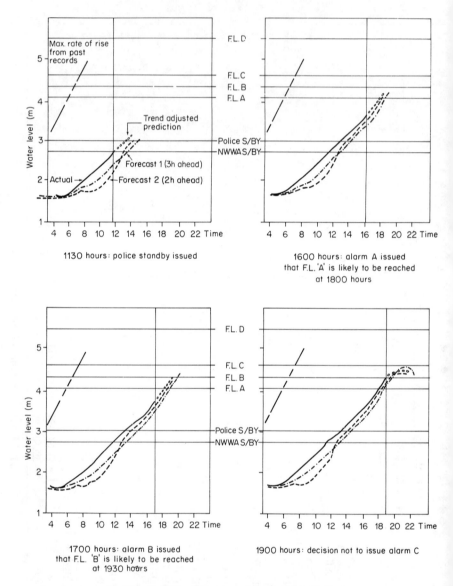

Figure 8.5. Progress of typical forecast

for some flood risk zones. Figure 8.5 shows the progress of a typical plot where a 3 h forecast is available.

Quite apart from models based on areal rainfall measured by radar, much is gained by having a radar display of current rainfall with the facility to play back the last eight pictures (which are at 15 min intervals). Duty officers can determine quickly the flood risk zones which might be affected by heavy rain, and it is most useful to discuss the weather pattern with the forecasters at the Manchester Meteorological Office who have a similar radar display. Further, the display is very useful in determining when rainfall stops; peak flow time for a flood risk zone may then be deduced, adding to the data produced by a model prediction.

Having determined from radar the broad pattern of potential flooding, the duty officer is able to display (in graphical form on the VDUs) historical and forecast data for any area that might flood. From this information he is able to determine the flood risk zones for which he needs to plot river levels and forecasts. It is worth noting that in spite of the facility to display levels on the VDUs, the intimate knowledge gained by manual plotting ensures that the best subjective adjustments are made to forecasts produced by models. Further, plots at all affected flood risk zones are immediately available, rather than the limited number (usually three or four) that may be displayed on a VDU at any one time.

When high tidal surges are predicted, in addition to monitoring tidal levels and subjectively correcting tidal forecasts, it is necessary to consider the combined effects of the high tide and river flows (predicted by routing and rainfall-runoff models) for estuarial flood risk zones.

Dissemination of forecasts

As soon as it is forecast that a 'police standby' level is likely to be reached then the appropriate county police are advised. This is usually a standard telex message issued from the flood forecasting room, but a telephone call from the duty officer's home may be necessary in a very rapidly changing situation. When it is forecast that various flood levels will be reached further standard telex messages are issued to the country police who in turn carry out duties at each stage in accordance with the flood warning schedule. Even in cases where only short period—say 2 h—forecasts are possible, issuing standby messages to the police at an earlier time puts them into a position where they, the local authorities and emergency services, are able to react very quickly in the event of actual flood levels being forecast. Public warnings are only issued when flood (not standby) levels are forecast.

NWWA operations staff, in addition to police and local authorities, are of course kept fully advised of all river and tidal forecasts so that they can take appropriate action. This action would include the monitoring of automatically operated sluice gates to flood storage reservoirs. Figure 8.6 shows diagrammatically the way that information is received, processed and disseminated.

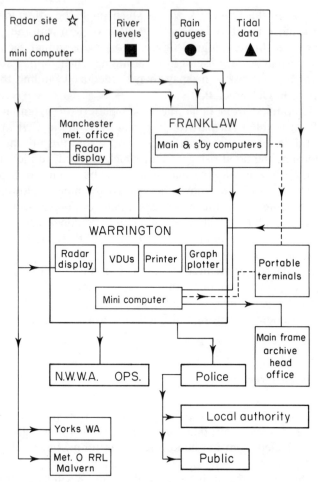

Figure 8.6. Flow diagram for data and forecast information

Duty rota and training

With present forecasting techniques it is possible to give flooding forecasts
for more than 20 flood risk zones. With developments in modelling procedures and with the advent of quantitative short-term rainfall forecasts from
the Meteorological Office, it will be possible to give useful forecasts for
many other flood risk zones. About seven teams, each of two duty officers,
will be on standby, operating on the basis of 1 week on duty and 6 weeks
off duty. In the event of a flood alarm, one team will initially operate the
flood warning procedures, supplemented by other teams if affected flood
risk zones become widespread and as 24 h manning becomes necessary.

All duty officers require training in forecasting procedures. To this end training sessions have been devised. Typically three flood risk zones are considered with hypothetical storms (but very closely related to actual storms) covering a 24 h period on the catchment areas. Reports equivalent to those which would be available from the computer are prepared for each $\frac{1}{2}$ h of the 24 h period. During training sessions each duty officer is given the messages at approximately 5 min intervals (i.e. speeding up time by a factor of six) and officers are expected to produce forecasts and prepare appropriate telex messages. In a real event the relative calm of a classroom environment does not prevail. Duty officers have to deal with many telephone calls, although administrative staff filter the calls leaving the duty officer to handle only those relating to flood forecasts. Further, one duty officer (or more as the number of affected flood risk zones increases) will be preparing forecasts, whilst the second duty officer deals with telephone calls. The exercises are well worthwhile, and give duty officers a good deal more confidence in their ability to handle flood warning procedures, particularly for new duty officers or those who have not had to deal with a real event for some time in their normal tours of duty.

Flooding from sewers

At present, available models and flood warning procedures cover areas where the flooding is from major rivers or from sewers which flood as a result of high river levels. In due course, when quantitative rainfall forecasts are available and where methods can be developed to predict direct flooding from sewers, then urban areas subject to this type of flooding can be incorporated in flood warning schemes.

DEVELOPMENTS BY OTHER BODIES

Other water authorities are using data for flood warning interrogating out-stations automatically by a computer using the telephone switched network. Further, they are examining the viability of transferring the data to archive. In general data loggers are used at outstations, and loss of data through faulty communications is rectified at a subsequent interrogation.

The Committee for Tidal Gauges, which is responsible for maintaining tidal gauges around the British Isles, is also implementing an automatic system for interrogating these gauges. Data from some gauges are used in real-time flood warning by the East Coast Storm Tide Warning Service. The existing data gathering system for the Tide Warning Service relies on dedicated telephone lines connecting tidal gauges on the east coast to repeaters in the Storm Tide Warning Service Office in Bracknell. The repeater equipment is now reaching the end of its useful life and is being replaced by interrogable data loggers which have been designed and developed by the Institute of Oceanographic Sciences at Bidston. The outstations are interrogated via the telephone switched network from a central computer at Bracknell. The computer may be set to carry out such interrogations at regular intervals from 24 h down to 15 min.

Also, for archiving data, a further computer at the Institute of Oceanographic Sciences at Bidston Observatory will automatically interrogate the out-stations once each day between midnight and 6 a.m. (when any number of UK calls can be made for a fixed charge). This system will replace the present method of collecting data by chart which has to be digitized to get the information into a computer. In due course data from all Grade A tidal gauges will be obtained in this way.

A system which has not been exploited to any degree in the UK, but which has been used extensively in the USA, is that of satellite communications. In the system, outstations (usually referred to as data collection platforms (DCPs) in satellite communications literature) are set up to read and store appropriate data (rainfall, river level, etc.). Each outstation then transmits the information in blocks at predetermined time intervals (usually between 1 and 24 h), synchronized with specific time slots on a satellite (one of the Meteosat series), which then relays the data to either the Meteosat control station at Darmstadt in West Germany for onward transmission to the user or to the user's base station where it is picked up by a 1.5 m diameter dish antenna. Any outstation can be designed so that in the event of alarm conditions being reached in any of the observed parameters it will send an immediate message to the base station advising the alarm condition and will, at the same time, increase its transmission rate (for example, from 24 h to 1 h or less if necessary) for sending data, the new transmission rate being held until the outstation goes out of alarm.

Collection of data by satellite has two main advantages. Firstly, outstations can operate in very remote areas where communications by normal radio transmission or by telephone line would be very expensive in capital costs and secondly, as only a very small aerial is required at the outstation it is much more acceptable in environmentally sensitive areas than the large mast often required by normal radio systems. The cost of a satellite system, whilst more than that of normal radio equipment, could be quite acceptable in certain circumstances. Typical costs for transmitting equipment at a single outstation with normal and alarm channels with the data being sent to Darmstadt are in the order of £5000 (1985 prices). A base station receiver which can serve several outstations and pick up data locally, rather than through Darmstadt, is about £15 000.

A licence has to be obtained to use satellite communications but there is normally no charge for transmitting hydrometric data between a DCP and the user's own receiver, although there would be a charge for transmission of data between Darmstadt and users who do not have their own receivers.

SUMMARY

The use of the regional communications scheme, with its capacity for rapid data transmission and its facility for running models, has improved very substantially the flood forecasting and flood warning capabilities of the North-West Water Authority. In particular, the following advantages are noted in comparison with the former system, which relied upon manually interrogable rain and river level gauges:

1. Formerly, instruments were only interrogated on a 24 h basis when conditions were dry and weather forecasts were good. Local isolated storms could easily be missed even though some critical areas had alarm-type gauges. Data now gathered at 15 min intervals from radar, and rain gauges enable the computer to identify such storms and raise appropriate alarms.
2. No indication was available if interrogable gauges broke down between the times that they were interrogated. There could therefore be a breakdown for as long as 24 h before the fault was known. With gauges now automatically interrogated at 15 min intervals faults are much more readily recognized.
3. When a duty officer received a heavy-rain warning from the Meteorological Office it was necessary for him to manually interrogate a number of gauges and manually produce some river level forecasts before he could make a decision on the seriousness of the situation. This could take between $\frac{1}{2}$ h and 1 h the information is now available in a few minutes from either office-based or portable computer terminals.

Experience has indicated that the RCS system is very robust, and it operates very well in adverse weather conditions when reliability is of utmost importance. Nevertheless, it is still felt prudent to leave telephone-interrogable gauges to give back-up at the most critical stations. It is significant that to date these have not been needed in a flood warning situation. The RCS system is not a cheap system. Indeed it is likely that if flood warning had been its only use, economic justification for its installation would have been impossible, and a system using the switched telephone network would have been considered. However, as flood warning was only one function in the total use of the system, economic viability was readily demonstrated.

The system is very flexible in that more outstations can be added and other forms of data can be handled; for example, when quantitative rainfall forecast data are available from the Meteorological Office, transmission into the RCS computer should be possible so that data can then be entered directly into rainfall-runoff models. The facility to be able to interrogate the RCS computer from home has also proved extremely valuable in allowing the duty officer to get both a detailed and general impression of any potential flooding situation within a few minutes of any initial alert.

There is no doubt that communications developments are revolutionizing the collection of real-time and non-real-time data; but it is perhaps appropriate that this chapter should end with a reminder that hydrometric data are only as good as the 'sharp end' equipment that reads river levels and rainfall and converts the data into a form that can be transmitted. It is hoped that equal development in these fields will match the improvements in data logging and communications.

Part III
MODELLING RUNOFF USING RADAR DATA

Weather Radar and Flood Forecasting
Edited by V.K. Collinge and C. Kirby
© 1987 John Wiley & Sons Ltd.

CHAPTER 9

UK Flood Forecasting in the 1980s[1]

D. W. REED

INTRODUCTION

The feasibility and success of a flood warning scheme depend on many factors: data acquisition and procedures for disseminating warnings are particularly important. However, it is in the development and application of flood forecasting methods that the hydrologist's skills are most relevant. It is convenient to distinguish between those methods based on flood routing and those based on rainfall-runoff. Although simple correlation techniques continue to find application in routing flood peaks, there is now considerable use of flood routing models such as the Muskingum–Cunge and variable parameter Muskingum–Cunge methods. A wide range of rainfall-runoff models have been developed for flood forecasting in the UK. These are discussed in this chapter under four groupings: unit hydrograph techniques, transfer function methods, conceptual models, and non-linear storage models, with particular mention of non-linear storage model applications.

This review highlights the special character of real-time applications of hydrological models and considers in detail the problem of how best to correct flood forecasts by reference to telemetered flow measurements. Three approaches are distinguished: error prediction, state-updating and parameter-updating. One general conclusion reached is that whereas standardization of methods is desirable in hydrological designs such as water resource assessment and flood estimation, the diversity of UK flood forecasting techniques in the 1980s is seen more as a strength than a weakness.

---

[1] This chapter is based on a review of British flood forecasting practice (Reed, 1984).

THE ROLE OF FLOOD WARNING

Flood warning can sometimes be considered as an alternative to flood alleviation as a means of reducing risk to life and property, albeit a less effective one. Even if the flood forecast is correct, the dissemination of the warning swift and thorough, and the public response prompt and effective, there remains the fact that the warning does not eliminate flooding but merely allows some reduction in damage. Although flood warning is a permissive power rather than a duty, most authorities take the view that the public expect to receive reasonable warnings of inundation of property.

The three watchwords in flood forecasting are accuracy, reliability and timeliness. Accuracy is clearly important if the forecast is to form the basis of specific flood alerts. If the forecast of peak river level is substantially in error then a false alarm may be raised or, worse, the system may fail to warn. Reliability of a flood warning system is primarily concerned with instrumentation, telemetry and procedural matters. But however well designed these aspects are, it is inevitable that the forecasting model will have to live through periods of outstation malfunction. It is therefore desirable that the forecasting method should reliably cope with imperfect or missing data: by validation checks, appropriate default values, and 'slimmed-down' models.

Accuracy and reliability are fairly obvious requirements; accuracy suggests we need a 'good' model, reliability hints that simplicity and robustness may be important. But the need for timeliness gives flood forecasting a flavour of its own. If warnings are issued consistently late then the system is likely to be of little value, irrespective of the accuracy of modelling. A balance has to be struck between issuing a timely but potentially inaccurate forecast and the more cautious approach of compiling a good picture of the event before issuing an accurate but useless forecast. In British conditions (i.e. relatively short, rapidly responding river systems) the response time of the catchment often imposes an upper limit on the forecast 'lead time' that can realistically be provided. Anything from 3 to 6 h is generally useful, with limited benefits accruing from shorter warning times also (Chatterton *et al.*, 1979).

Shortcomings exposed in a major flood incident provide an obvious incentive to improved flood warning. It is generally possible to instigate new warning procedures fairly quickly perhaps as an interim measure until river improvement works are carried out or while improved flood forecasting methods are developed. Another strong influence is the enhanced scope for flood warning presented by technological developments, with a particularly striking example being weather radar. Such developments raise expectations and *ipso facto* create pressure for improved flood warning.

DATA ACQUISITION

Increasingly, computers are being used to control real-time data acquisition. Most authorities favour divisional or regional telemetry schemes where practical. Attention is generally focused on the recording and display of data—and on the reliability of data acquisition—with the implementation of flood forecasting programs following as circumstances allow. An alternative approach is to concentrate resources on a particular flood warning problem (Brunsdon and Sargent, 1982). Although rare as yet, this approach may gain ground in the general trend towards more personalized, distributed and dedicated computers.

Perhaps the most marked characteristic of computer-controlled telemetry systems is their capacity to generate vast quantities of data. Good management of data storage is essential if the system is to serve for data archiving as well as flood forecasting. Even in systems where data are retained only for short-term use, much thought is required in structuring the database (Evans, 1980).

Prior to placing newly telemetered observations in the database, some form of validation is generally carried out, because whereas an experienced eye can spot a rogue value instinctively, a computer has to rely on a checking formula. Validation of rain gauge readings is particularly difficult because of the acute variations in rainfall intensity that can occur both in space and time. Practical experience of computerized validation of telemetered data has been gained by several authorities although, as yet, it is a detail of system design which has not been widely publicized. Effective data validation extends the scope for computer surveillance of instrument failure.

With an efficient data collection and display system it may be sufficient to base flood warnings on present conditions rather than adopt forecasting techniques. This may be attractive where river level is monitored some distance upstream of the risk site. Of course the success of such an approach relies on a relationship existing between river conditions at the two sites but, by setting alarm levels from experience, it is possible to use the relationship without actually formulating a model of it.

Good presentation of rainfall and flow data can similarly allow reasoned flood warnings to be issued without recourse to formal methods of flood forecasting. In essence, the duty officer supplies the model by visually relating cause (heavy rainfall) to effect (rising river levels). Scope for successful application of subjective methods has been greatly enhanced by computer graphics developments. Whereas formerly deductions had to be made from simplified information, or time lost in drawing up graphs, it is now feasible in computerized systems to display updated hydrographs and hyetographs only moments after interrogation.

DIVERSITY OF METHODS

Very many methods of flood forecasting are in operational use in Britain. Whereas flood warning procedures often follow a divisional or regional pattern, there is much greater variety in the methods used to calculate forecasts. A few water authorities, notably Severn–Trent WA (Douglas and Dobson, Chapter 11 this volume), have evolved a regional strategy for flood forecasting but, in the main, methods are tailored to local requirements and preferences. Factors influencing the choice of flood forecasting method include:

1. the frequency and level of threat to life and property;
2. the availability of historic data;
3. outstation, telemetry and computer resources;
4. staff time and abilities, personal preferences;
5. the fulfilment of a dual purpose (e.g. a single forecasting model applicable to both flood and low flows).

But the dominant influence is likely to be the nature of the catchment.

Although small scale by world standards, the range of flood forecasting problems in Britain is nevertheless very wide. A flood routing approach is generally practical for large catchments but on more rapidly responding catchments some element of rainfall-runoff modelling may be required. A particular challenge is forecasting floods on very rapidly responding upland or urban catchments where quantitative rainfall forecasts may be required if existing short-period warnings are to be enhanced significantly.

SPECIAL CHARACTER OF REAL-TIME APPLICATION

The demand put on a rainfall-runoff or flood routing model for flood warning are rather different from those imposed in other applications, such as design flood estimation. In the first instance, the model should be capable of providing a reasonable forecast based on data available at the time of forecast, i.e. without recourse to additional information such as daily rain-gauge readings (or the time of peak flow!) that is available only in retrospect. This very obvious requirement is surprisingly easy to overlook when adapting flood estimation methods to flood forecasting.

In design flood estimation applications, the aim is to provide either a peak flow estimate or a stylized design hydrograph. In flood warning applications there is likely to be more emphasis on timings and the reproduction of distinctive shapes on the rising limb and crest segment of the hydrograph.

A further distinction is that flood forecasting on ungauged catchments is rarely considered. Application of generalized models to ungauged catchments is inherently inaccurate and, whereas a relatively uncertain flow estimate may be quite acceptable for engineering design purposes, for flood warning it may be better not to warn than to do so with great inaccuracy. In flood warning, some form of river level measurement at or near the risk site is highly desirable; otherwise it will be very difficult to interpret model forecasts.

This leads us to consider the most fundamental difference between flood forecasting and other applications: the use of downstream flow information available in real time. If, as is usual, the forecasts made refer to flow at a telemetered site, the opportunity exists to compare recent flow observations with the corresponding values simulated by the rainfall-runoff or flood routing model.

FLOOD ROUTING METHODS

On some of the longer river systems in Great Britain, satisfactory flood warnings can be based on an upstream gauging station. This approach is generally to be preferred where practicable. The alternative of basing forecasts on rainfall measurement and rainfall-runoff modelling is more prone to error. It is, of course, the river that floods and for that reason a measure of river flow is generally a better indicator of flood potential than is rainfall. By the same argument, flood routing methods can be of some use in floods arising from snowmelt, whereas rainfall-runoff methods require special consideration of snow accumulation and melt.

The upstream gauging station may be part of a general-purpose river gauging network or be specifically designed for flood warning. A well-defined flood rating is desirable though not essential. River level at the risk site is the crucial variable in terms of inundation, and hence warnings are usually based and issued in terms of river level. For the approach to be successful, the travel time of flood peaks from the upstream gauging station to the risk site needs to be long enough to allow an adequate period of warning to be given, say 4–6 h. But the gauging station must not be so far upstream that it is unrepresentative of flows to be expected at the risk site.

Fairly simple graphical methods based on correlation of upstream and downstream flows or river levels continue to be useful. However, if hydrograph shape and timings are important, or tributaries confound the problem, a flood routing model proper is called for. Flood routing models range from the simple to the highly complex, and most require computer implementation. The variable parameter Muskingum–Cunge (VPMC) method seems to offer an appropriate compromise between simpler but less realistic methods and more complicated hydraulic methods with large data and computing requirements. Some further development of VPMC for real-time use would be valuable. More complete solutions of the St Venant equations are appropriate for reaches subject to tidal or backwater influence.

RAINFALL-RUNOFF METHODS

A wide range of rainfall-runoff models can be considered for flood forecasting in British conditions. Unit hydrograph methods have a structure and familiarity that appeal to many engineering hydrologists although the approach is less suited to real-time forecasting than it is to design flood estimation.

Transfer function models are broadly equivalent to unit hydrograph methods but intrinsically better suited to real-time use. They are simple to implement and easily re-initialized but their calibration calls for not inconsiderable statistical expertise. A common weakness of transfer function and unit hydrograph methods is the lack of convincing guidance on rainfall separation techniques.

Conceptual models are generally rather cumbersome for real-time use. An appropriate choice of model structure calls for a relatively deep insight into those factors most relevant to runoff generation. Conceptual models are usually calibrated by numerical optimization of continuous sequences of flow and climate data. Proponents argue that it is necessary to have a physically sound model if the model is to perform reliably in extreme events. Sceptics reply that parameter values derived by optimization may be spurious and those fixed *a priori* may be subjective.

A BRITISH SPECIALITY: NON-LINEAR STORAGE MODELS

Non-linear storage models collectively form the most widely used rainfall-runoff method in British flood forecasting practice. The method encompasses as special cases the inflow–storage–outflow (ISO) function techniques (Lambert, 1972) and the isolated event model developed in the mid-1970s (NERC, 1975). The essential features of a non-linear storage model are illustrated in Figure 9.1. Rainfall (p) is lagged by a pure time delay (L), reduced to net rainfall (n) by application of a runoff proportion (ROP), before being routed through a non-linear storage (S) to produce the outflow or catchment runoff (q).

Several variants have been developed and can be characterized by the particular functional forms chosen for the runoff proportion (ROP) and the routing function, $dq/dS(q)$. For example, in the isolated event model, ROP is determined by the pre-event soil moisture deficit (SMD) and the routing function is proportional to the square root of the outflow. (See insets to Figure 9.1.)

The somewhat simpler ISO function techniques make no explicit allowance for rainfall losses, setting ROP = 1. Although restricting their use to relatively impermeable catchments, this omission is less damning than at first

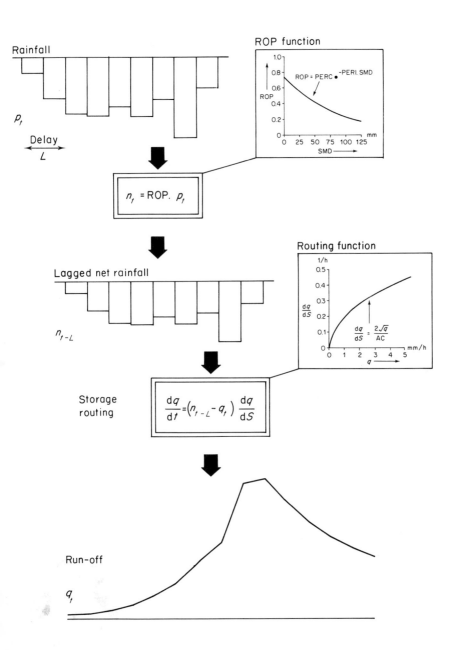

Figure 9.1. Non-linear storage model structure, with (inset) ROP and routing functions for the isolated event model variant

appears. Firstly, some indirect compensation is possible during model calibration. Secondly, if—as is usual—flow data are available in real time, it is
possible to re-initialize the solution of the storage routing as each newly
telemetered flow value is received. As will be seen shortly (Figure 9.3), it
is this feature, more than any, that endears non-linear storage models to the
flood forecaster.

Eyre and Crees (1984) have applied the isolated event model (IEM) to
flood forecasting on a rapidly responding catchment in north-west London.
The non-linear storage model applied in Forth River Purification Board's
Haddington flood warning system (Brunsdon and Sargent, 1982) is the
modified IEM, in which the observed pre-event flow determines the runoff
proportion applicable to the current event. Subsequent analysis of a relatively
permeable catchment in Lincolnshire (Reed, 1984) confirmed that pre-event
flow is a useful surrogate measure of catchment wetness in the context of
relatively simple rainfall-runoff models for flood forecasting.

ISO function models remain a notable component of the Welsh Water
Authority's Northern Division's flood forecasting armoury, both in the Dee
system (Lambert and Lowing, 1980)—the forerunner of comprehensive computerized flood warning in Britain—and in graphical form for lesser rivers
such as the Elwi. More recently, ISO function methods have found favour
in World Meteorological Organization projects in the Caribbean, through
their technology transfer programme HOMS (Lambert and Reed, 1986).

More advanced non-linear storage models have sought to represent translation effects by a time–area diagram rather than a simple pure time delay.
For example, the semi-distributed catchment model developed for the Bristol
Avon has this feature. The relevant Wessex Water report (Grimshaw and
Wong, 1980) is typical of many detailed but unvaunted flood forecast modelling investigations. Although formulated more as a regression method, the
rainfall-runoff model developed by North West Water to run automatically
and continuously on the authority's regional communications computer is a
final example of a non-linear storage model that has found practical application (Knowles, Chapter 10 in this volume).

REAL-TIME CORRECTION

How best to use telemetered flow data to correct a forecast is a central
theme in flood forecasting. Figure 9.2 illustrates the basic problem: the
model is yielding a flow forecast that telemetered observations show to be
inaccurate. Should the forecast projected by the model be trusted, or is it
better to try to correct the forecast in the light of recent model performance?

The question can arise in both flood routing and rainfall-runoff modelling
methods of flood forecasting. There are at least three distinct ways of
correcting a 'simulation mode' forecast by reference to telemetered flows:

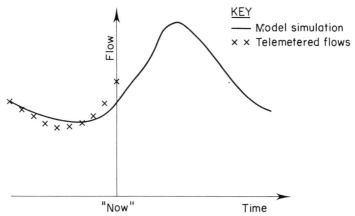

Figure 9.2. The need for real-time correction

1. *Error prediction*—in which recent discrepancies between simulated and telemetered flows are studied and a corrected forecast constructed by adding error predictions to the simulation mode forecast.
2. *State updating*—in which the catchment outflow (or some other observable quantity) acts as a state variable so that a telemetered observation can be used to update the state of the model (and hence its forecasts) directly.
3. *Parameter updating*—in which one or more of the model parameters are varied in the light of recent model performance.

Error prediction

The most obvious approach to the problem posed by Figure 9.1 is to accept that a discrepancy exists between the model forecasts and the flow observations, and to try to anticipate how this is likely to develop in the near future (i.e. within the forecast lead time). The aim of the approach is to reconcile what the model is telling us *will* happen with what the telemetered flow observations say *is* happening. The reconciliation is most easily done graphically, 'blending' the two pieces of information together. The essence of the technique is to predict the model error at future time steps from the model error noted in the recent past—hence the term 'error prediction'.

State updating

Figure 9.3 illustrates the state-updating approach to real-time correction, using as an example the isolated event model (IEM). Here the latest tele-

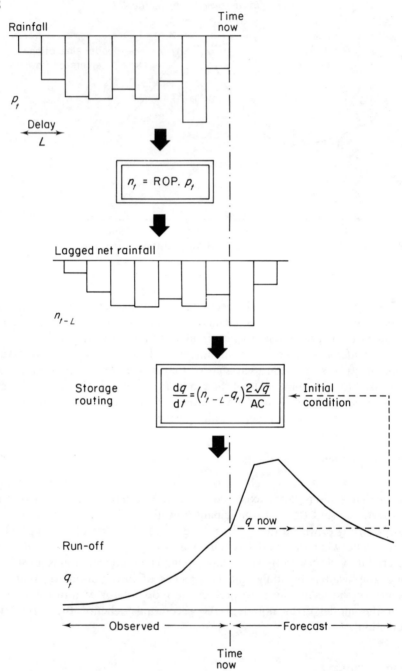

Figure 9.3. IEM structure for real-time forecasting, illustrating re-initalization (state-updating) by reference to telemetered flow

metered flow is used to update the 'state' (i.e. content) of the model's non-linear reservoir. The effect is to re-initialize the model forecast so that it tallies with the observed flow at time 'now'. The broad philosophy behind the state-updating approach is that the model can be trusted for short-term flow forecasting provided that it is always starting from the right place.

Parameter updating

The third approach to real-time correction is more drastic. One or more of the model parameters are adjusted in some systematic manner until model performance is more in keeping with recent flow observations. This is the parameter-updating approach, sometimes referred to as parameter adaptation or, simply, adaptive forecasting. When using the approach it is necessary to ask several questions. Which parameters should be modified? And by what criteria? The possibilities are, of course, legion. Usually no more than two parameters are adjusted, perhaps a 'volume' parameter and a 'timing' parameter. Often the adjustment is made to minimize some sort of least-squares criterion between recent model forecasts and telemetered flows. This raises further questions. Should the criterion be weighted to give more emphasis to very recent flow discrepancies? How frequently should parameter updating be attempted?

Which approach?

Most real-time correction undertaken in British flood forecasting practice is of the error prediction type. Graphical methods are the most common, with the forecaster deducing a suitable compromise forecast when presented with model forecasts and telemetered flows that disagree. However, automating what the forecaster's eye can judge instinctively is not particularly easy. The state-updating technique is used by a number of flood warning authorities, primarily in connection with ISO function models. Parameter-updating methods of real-time flow forecasting are, in Britain, as yet largely confined to research use.

If time and facilities permit the flood duty officer to intervene, visual comparison of model forecasts and telemetered flows is the simplest and probably most reliable means of real-time correction. Automatic methods of correcting forecasts—whether they be of the state-updating, parameter-updating or error prediction type—may confuse a duty officer. If unable to understand the way in which the model forecast is changing, the officer may be left with only two options: to reject the forecast or to rely on it fully. Too much reliance on automatic methods may engender a false sense of security. In most cases it is therefore desirable that the uncorrected (i.e. simulation mode) forecast should be available for comparison.

In choosing an appropriate method of real-time correction it may help to consider the likely cause of errors in the model forecast. If a particular part of the model is known to be important but error-prone, then a parameter-updating technique may be appropriate. In contrast, a state-updating technique may be indicated if the input data or the initialization of the model is considered suspect. If little is known of the particular weakness of the forecasting model, an error prediction method is probably the safest choice.

Error prediction methods based on time series analysis can be used to model the residual error of any model forecast; they provide a natural way to formalize visual correction methods (Box and Jenkins, 1970; Moore, 1982). A potential weakness of automatic methods is that the correction may be led astray by a timing discrepancy between forecast and telemetered flows.

MODELLING OF LARGE RIVER SYSTEMS

Some of the larger river systems in Britain merit a fairly comprehensive approach to flood forecasting. For example, the Severn and Trent river basins are represented by a combination of subcatchment rainfall-runoff models and main river flood routing (Douglas and Dobson, Chapter 11 in this volume). This semi-distributed approach to catchment modelling has several advantages: it allows flow forecasts to be generated for several sites, extends the lead time attainable by flood routing alone, and permits direct account to be taken of spatial variations in rainfall.

SNOWMELT FORECASTING

Measuring and modelling snowpack behaviour presents a difficult problem for flood forecasters in Britain. Snowmelt forecasting is relatively well met in subarctic conditions by empirical methods such as the 'degree–day' method. But in much of Britain, significant snow accumulations are relatively infrequent, may be spatially quite varied and short-lived, and may (or may not) be accompanied by a frozen subsoil. These factors conspire to make snowmelt forecasting a difficult problem.

Not surprisingly, most flood warning authorities rely on 'experience' methods of forecasting snowmelt runoff. Whether complicated physics-based models of snowpack behaviour (Morris, 1983) offer significant benefits to the flood forecaster over empirical methods is as yet unclear. In snowmelt forecasting the opportunity undoubtedly exists to exploit sophisticated scientific techniques—such as remote sensing and physics-based models—to help resolve an acknowledged problem. Perhaps the question to be answered is whether the likely benefits justify the resources needed to implement the science.

WHAT NEXT?

As stated earlier, the diversity of UK flood forecasting techniques now available in the 1980s is seen as an advantage over the trend towards standardization of methods which is more desirable in hydrological designs such as water resource assessment and flood estimation. The distinction arises because in flood forecasting it is possible to assess method performance after each event for which it serves operationally. Perhaps what is needed now is a more open discussion of the successes and failures of current flood warning systems.

In this chapter little mention has been made of weather radar. However, it is clear that the potential of radar to enhance flood forecasting in UK conditions is twofold: through direct use of rainfall estimates in hydrological models and, indirectly, through improved short-term rainfall forecasting. One goal—as yet unattempted—would be to provide a 'catch-all' method for detecting freak rainfalls and to anticipate resultant flooding through the use of digital terrain models. Perhaps the most challenging aspect of such a project would be the development of general procedures to alert communities for which no regular flood risk is perceived.

No less fanciful is the concept of adopting a probabilistic approach to flood warning (Creutin and Obled, 1980). Both rainfall forecasting and rainfall-runoff modelling are uncertain activities. In a probabilistic approach the output of a forecasting calculation could be a statement of the likelihood of a given critical river level being exceeded within the forecast period. This approach would expose the level of uncertainty in flood forecasting and might provide a framework for subjective factors—such as the public judgement of forecasting errors—to influence the decision whether to issue a warning.

ACKNOWLEDGEMENTS

The review was carried out as part of a research project funded by the Flood Protection Commission of the Ministry of Agriculture, Fisheries and Food. The co-operation of flood warning authorities is gratefully acknowledged.

REFERENCES

Box, G. E. P., and Jenkins, G. M. (1970). *Time Series Analysis, Forecasting and Control.* Holden-Day, San Francisco.
Brunsdon, G. P., and Sargent, R. J. (1982). The Haddington flood warning system. *Proc. Exeter Symposium on Advances in Hydrometry*, IASH publ. no. 134, pp. 257–72.
Chatterton, J. B., Pirt, J., and Wood, T. R. (1979). The benefits of flood forecasting. *J. Inst. Water Engrs Sci.* **33**, 237–52.
Creutin, J. D., and Obled, Ch. (1980). Modelling spatial and temporal characteristics

of rainfall as input to a flood forecasting model. *Proc. Oxford Symposium on Hydrological Forecasting*, IAHS publ. no. 129, pp. 41–9.

Evans, G. P. (1980). Data handling aspects of real-time flow forecasting. Paper at IH/WRC colloquium on real-time flow forecasting, Medmenham.

Eyre, W. S., and Crees, M. A. (1984). Real-time application of the Isolated Event rainfall-runoff model. *J. Inst. Water Engrs Sci.*, **38**, 70–8.

Grimshaw, D. E., and Wong, T. (1980). Bristol Frome investigation: calibration and refinement of catchment model. Wessex Water Bristol Avon Division report no. PD/022/1.

Lambert, A. O. (1972). Catchment models based on ISO-functions. *J. Inst. Water Engrs Sci.*, **26**, 413–22.

Lambert, A. O. and Lowing, M. J. (1980). Flow forecasting and control on the River Dee. *Proceedings of the Oxford Symposium on Hydrological Forecasting*. IAHS Publication No. 129, pp. 525–534.

Lambert, A. O., and Reed, D. W. (1986). Rainfall/runoff and channel routing using the ISO-function method. Report forming HOMS component J12.1.08, Institute of Hydrology/World Meteorological Organization.

Moore, R. J. (1982). Transfer functions, noise predictors, and the forecasting of flood events in real time. In Singh, V. P. (ed.): *Statistical Analysis of Rainfall and Runoff*, Water Resources Publ., Colorado, pp. 229–50.

Morris, E. M. (1983). Modelling the flow of mass and energy within a snowpack for hydrological forecasting. *Ann. Glaciol.*, **4**, 198–203.

Natural Environment Research Council (1975). Conceptual catchment modelling of isolated storm events. Vol. I, Section 7.3, *Flood Studies Report*, NERC.

Reed, D. W. (1984). A review of British flood forecasting practice. *Report* no. 90, Inst. Hydrol., Wallingford.

Weather Radar and Flood Forecasting
Edited by V.K. Collinge and C. Kirby
© 1987 John Wiley & Sons Ltd.

CHAPTER 10

Flood Forecasting Hydrology in North West Water

J. M. KNOWLES

INTRODUCTION

Before the North West Water's Regional Communications Scheme (RCS) was commissioned, the flood warning systems depended upon gathering data by manual interrogation, by telephone, of rainfall and river level stations. Monitoring was generally triggered by the Meteorological Office giving forecasts of rainfall in quantities above predefined limits, although certain stations had automatic alarm instruments. For flood risk areas with longer response times (Figure 8.1), such as those on the Rivers Eden, Lune, Irwell and Mersey, forecasts were based on simple regression models using up to three upstream water levels. Forecasts for short-response catchment areas, like the River Cocker and upper Eden, were based on unit hydrograph rainfall-runoff models. The limitations of the manual system of collecting and processing data meant that forecast times were short and there was clearly much scope for further development.

The completion of the RCS and the weather radar system (Figure 8.2) made available a wealth of data as follows:

1. Data from river level stations and rain gauges within the RCS network, which are automatically obtained every 15 min, validated and stored on the RCS computer at Franklaw for 48 h before being overwritten.
2. Data from the Hameldon Hill weather radar, which are automatically obtained every 15 min. These data are available both as a display on a colour monitor and as numerical rainfall values over predefined catchment areas. The catchment data are passed to the RCS telemetry system and are then dealt with in a similar manner to gauged rainfall.

It is expected that at some time in the future a further source of data will be forecast rainfall provided by the Meteorological Office FRONTIERS system.

Hydrological models incorporated in the RCS computer allow the above data to be used for forecasting, and alarms can be generated automatically from any of the collected or forecast values. This chapter describes the search for suitable models, and details some results obtained from application of the preferred models.

TYPES OF RAINFALL RUNOFF MODELS CONSIDERED

Unit hydrograph (UH) models

Unit hydrograph theory, widely employed in design, was the basis of a number of studies aimed at developing suitable models for real-time application. The principle depends on there being a linear relationship between 'effective' rainfall and runoff.

Lancaster University (under a contract from Water Research Centre) developed a method for estimating storm runoff volume and rainfall losses, with encouraging results. However, further work would be required before this research could be applied in real-time use (O'Donnell and Groves, 1979). North West Water, in initial studies with UH models on a number of rivers (North West Water, 1977–80) found that quite good results could be achieved by relating catchment losses to the base flow of the river at the start of an event. In view of the amount of computation and data handling involved in storing and retrieving forecasts, these models have not yet been used in the North West Water real-time RCS system. Three of the UH models, however, were used by the Water Research Centre in a real-time forecasting system for the Mersey Basin, which is described more fully in a later section.

Although UH methods afford a reputable method of runoff forecasting, even for ungauged catchments, the problems of 'effective' rainfall estimation, and of dealing with breaks in data in a real-time situation, led to a search for other less demanding models with the added facility of adjusting the forecast as an event proceeds.

Models with automatic correction facilities

The Institute of Hydrology—in work funded by the Water Research Centre and the Ministry of Agriculture, Fisheries and Food—evaluated a variety of flow forecasting methods and developed a transfer function noise (TFN) model, which gives forecast flows based on recently observed flows and rainfall amounts (Moore and O'Connell, 1978, 1980). In this model a variable term is introduced which corrects the forecasts by reference to observed errors in the preceding forecasts. This correction, based on an autoregressive moving average (ARMA) of the observed errors, improves the forecasts and allows relatively quick recovery of the model following loss of data.

Further applications of an ARMA-type model were developed at Birmingham University under a CASE Studentship supported by the North-West Weather Radar Consortium (Cluckie and Owens, Chapter 12 in this volume). The model uses radar-derived catchment rainfall and observed flows with an updating term to adjust for variations in the percentage runoff. The method will also allow the use of forecast rainfall which will become available from the Meteorological Office FRONTIERS project.

Inflow storage outflow (ISO) models

North West Water has developed a rainfall/runoff model to run automatically and continuously on the Authority's RCS system (North West Water, 1984). The model is a variation of an earlier study (Lambert, 1972) where the assumption was made that the catchment area had a natural storage which controlled the rate of runoff ('outflow') and which was intermittently replenished by rainfall ('inflow'). In the North West Water model, which may be represented as:

$$Q_{T+1} = aQ_T + bR_T + Q_B$$

each forecast (Q_{T+1}) is a function of the previous forecast (Q_T), the rainfall since the previous forecast (R_T), and the measured base flow at the start of the event (Q_B). In the equation a is a recession constant and b is a parameter related to effective rainfall; these parameters and a suitable time delay (lead time) are determined and optimized from a number of events to give the best fit for flood flows. Initially the base flow term was simply the measured flow at the start of an event, but the model is being improved by making the term variable throughout the event. Further automatic correction procedures based on measured river levels are being investigated.

TYPES OF ROUTING MODELS CONSIDERED

Peak level correlation method

In this method, peak levels recorded at upstream river level stations are correlated with peak levels recorded at downstream stations near to flood risk zones (North-West Water, 1979–80). The times of travel of floods are determined from a number of events. On river systems with one or two major tributaries, multiple correlation equations have been developed. In most cases these models have performed well for forecasting flood flow but in common with all routing models their use is limited to larger catchment areas where they can give adequate lead times. Advantages of this type of model are that few parameters are needed, and they can be operated continuously and automatically on computer systems similar to that in the North West Water's RCS system. Problems with the method are that it relies on the observation of a number of high river levels to determine the correlation equations, and does not work too well on rivers with many tributaries between reporting stations. Updating methods to overcome the fixed lead time are being developed.

Muskingum–Cunge method

In this method the river is divided into separate reaches and the channel geometry and speed of flow are calculated for each reach. Upstream observed flows are routed through each reach in turn to give an estimate of the flow at the flood risk zone. Advantages of this model are that it deals with tributaries better, and requires fewer high river flow observations for its development than do peak level correlation methods. Disadvantages are that it does not model the hydrograph shape well, and does not take into account the changing geometry and speed of flow as the discharge alters.

The Water Research Centre, however, showed how this method could be developed and applied in real-time operation (Cameron, 1980). Their method uses an updating procedure with a Kalman filtering technique operating on the differences between the forecast and the observed flow to correct the forecast. Satisfactory forecasting performance was obtained for test cases where a good estimate of tributary/lateral inflow could be provided.

The Hydraulics Research Station applied their river catchment model 'FLOUT' to four rivers in the north-west, to assess its suitability for operating in the real-time system (Hydraulics Research Station, 1981). The model is a development of the Muskingum–Cunge method taking into account the variation of channel characteristics and speed of flow with discharge. Although giving quite good simulation of the whole hydrograph the computer program necessary to operate the model is relatively large and appreciable program amendments would be necessary before it could be used for real-time forecasts.

TYPES OF COMBINED MODELS CONSIDERED

In any of the routing models the upstream flows, including tributary flows, may be synthesized using rainfall-runoff models. This extends the lead time which would be available if only observed upstream values were used. Indeed, the FLOUT model was designed at the outset to include synthesized inflows.

Pilot flood prediction project

The North West Water Authority and the Water Research Centre co-operated in developing a computer-based forecasting system which can use telemetered observations of rainfall, measured by gauge or radar, and river flows together with quantitative rainfall forecasts (Cameron and Evans, 1980; North-West Water, 1980). The system is capable of 'patching in' any missing telemetered values by using data based on the nearest available stations. The computer program is sufficiently flexible to accept any combination of the routing and rainfall-runoff models described in this chapter.

In practice the model has been tested on the River Mersey catchment area where two raingauges, four radar subcatchment areas and four river level stations were used. Unit hydrograph models developed by North West Water were used in the project. These methods could operate with forecast rainfall of various intensity patterns. With appropriate forecast rainfall, river level predictions for up to about 12 h could be produced.

The advantage of the system is its ability to accept the most suitable models for any catchment area. The system has been designed to run interactively and could be very useful in complex catchment areas. However, a disadvantage of the system is that it places great demands on the operator, thus restricting its use operationally to one or two flood risk areas.

UNGAUGED CATCHMENT AREAS

Some flood risk zones have no telemetered or historic flow information but it was felt that forecasts could be made using the radar rainfall in a suitable model. The Institute of Hydrology reviewed the regression equations developed in the Flood Studies Report (NERC, 1975) with specific reference to the catchments in the North West. Their report (Boorman, 1980) showed that there was insufficient evidence to justify any change in the method of estimating the UH model parameters from readily available catchment characteristics. However, it was stressed that every effort should be made to get even a few river level observations at the flood risk zone, because of the uncertainty of this method of flood estimation.

OPERATIONAL USE OF MODELS

Factors which have been taken into consideration when choosing a model for operational use are:

1. The accuracy of the forecast.
2. The time available between the preparation of the forecast and the flooding taking place (lead time).
3. The manpower and computer resources required to operate the model under operational conditions.

For operational flood warning purposes, models can be considered to fall into two categories:

1. Models that can operate continuously and automatically in a system such as North West Water's RCS system and which would produce river level forecasts for a specific, relatively short, time ahead (ISO, peak correlation).
2. More sophisticated models requiring greater computer capacity and which could, with suitable quantitative forecasts of rainfall, produce forecasts with much longer lead times (TFN, ARMA, Muskingum–Cunge, FLOUT, pilot project).

A disadvantage with the present models running on North West Water's RCS system is that they have no facilities for automatic real-time correction, although updating procedures are being investigated. This disadvantage is overcome to a large extent, however, by the way in which the models are used. Available forecasts for a flood risk zone are plotted and the observed river level in the zone may then be extrapolated by reference to the forecast trends. The operational convenience and the ease with which this correction procedure may be applied has led to the choice of automatic, continuously operating models for use in North West Water's current flood warning procedure. The system operates on a 15 min data collection and processing interval, and forecasts are produced automatically on this basis. Calibrated radar rainfall data have been used in the rainfall-runoff models wherever appropriate. It is recognized that these models may not work in some of the more complicated catchment areas that still have to be considered, and that more sophisticated models may be appropriate, particularly when quantitative rainfall forecasts become available.

Figure 10.1 shows an example of forecast for the Ribchester flood risk zone on the River Ribble. The river level telemetry station serving this zone is just downstream of the confluence of two major tributaries. To take advantage of the distributed rainfall data available from radar, the total catchment area of over 1000 km^2 was split into three, covering the main river and the two tributaries, each with catchment areas of 260–460 km^2. Rainfall-runoff models were developed for each area using calibrated radar rainfall with the NWW modified ISO model derived from historic storm events. River level measurements are not available in real time from these

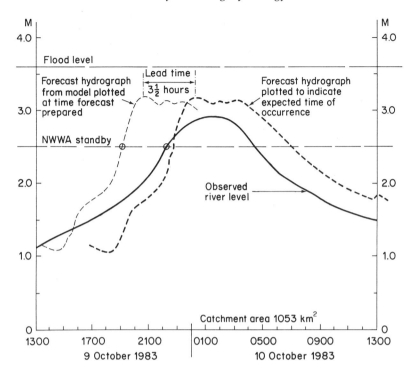

Figure 10.1. Forecast for the River Ribble

areas so no base flow correction parameters are used. The separate forecast flows were combined in a multiple regression peak flow/level model to produce forecast levels for the site, the lead time at high flows being approximately $3\frac{1}{2}$ h. This example shows that the standby alarm occurred some $3\frac{1}{4}$ h before the actual water level reached that level, and some $6\frac{3}{4}$ h before the peak was reached.

Further examples are provided by forecasts for the flood risk area at St Michaels on the River Wyre, shown in Figure 10.2. Both used a calibrated radar rainfall for the whole catchment area of 275 km^2 in the NWW ISO rainfall-runoff model. The forecast lead time is approximately 4 h. In both cases no base flow parameters were used, though these have subsequently been included to improve the fit of the forecast hydrographs.

The last example is for the River Darwen (catchment area 40 km^2) where the future availability of FRONTIERS data will significantly improve the current lead time of 1 h. Similar use of the NWW ISO model is made, but in this case incorporating a base flow parameter.

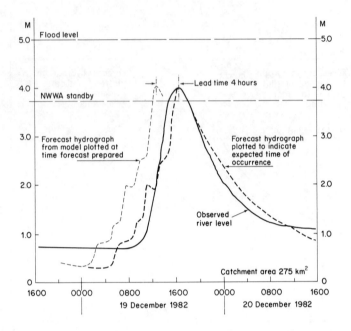

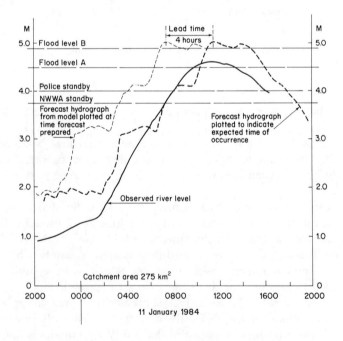

Figure 10.2.   Forecasts for the River Wyre

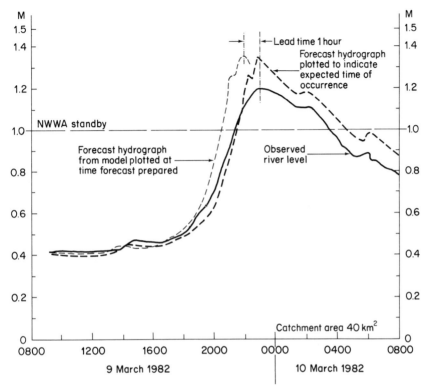

Figure 10.3. Radar forecast for the River Darwen

CONCLUSIONS

In general peak levels can be under- or over-estimated by the current model but the times for critical levels being reached are usually forecast very well. Further development to improve the forecasting capabilities of the system include more parameters in the models and the adoption of automatic updating along the lines of the work carried out at the University of Birmingham (Chapter 12, this volume).

REFERENCES

Boorman, D. B. (1980). A review of the flood studies rainfall/runoff model prediction equations for North West England. Institute of Hydrology report to MAFF (for NWRP). December 1980.

Cameron, R. J. (1980). An updating version of the Muskingum–Cunge flow routine technique. *Proc. Oxford Symposium*, IAHS publ. no. 129, pp. 381–87.

Cameron, R. J., and Evans, G. P. (1980). North West Radar Project Report Hydrological Forecasting Software. Water Research Centre Report 22-M. November 1980.

Hydraulics Research Station (1981). Flood routing in the rivers Ribble, Mersey, Weaver and Eden. Report no. DE52.

Lambert, A. O. (1972). Catchment models based on ISO-functions. *J. Inst. Water Engrs Sci.* **26**(8), 413–22.

Moore, R. J., and O'Connell, P. E. Real-time forecasting of flood events using transfer function noise models. Part 1—Institute of Hydrology, October 1978; Part 2—Institute of Hydrology, May 1980.

NERC (1975). *Flood Studies Report.* 5 vols.

North-West Water (1980). Pilot Flood Prediction Project, NWW Software Documentation. NWW Rivers Division. Radar Project Report. Report no. 30. September.

North-West Water (1977–80). NWW Rivers Division: Radar Project Reports: Rainfall–run-off models using unit hydrograph techniques.

North-West Water (1979–80). NWW Rivers Division: Radar Project Reports: Peak level correlation model.

North-West Water (1984). NWW Rivers Division: Radar Project Report: Rainfall/run-off models using modified ISO. (In preparation).

O'Donnell, T., and Groves, J. E. (1979). *A hydrological investigation of storm run-off volumes in the North West*, Vols 1 and 2. Report for Water Research Centre by University of Lancaster. October.

Weather Radar and Flood Forecasting
Edited by V.K. Collinge and C. Kirby
© 1987 John Wiley & Sons Ltd.

CHAPTER 11

Real-time Flood Forecasting in Diverse Drainage Basins

J. R. Douglas and C. Dobson

INTRODUCTION

Severn–Trent Water Authority provides a flood forecasting and warning service both for urban areas and agricultural floodplains. Larger urban areas often have significant flood protection, whereas agricultural land may be unprotected or have only low-level defences. The design of a flood forecasting system which can provide satisfactory warnings for these diverse flooding regimes must take account of the basin's physical characteristics. Both the Severn and Trent basins (Figure 11.1) are each in excess of 10 000 km² with a dendritic drainage network around a major river over 200 km long. These features mean that the emphasis is on the use of flood routing models as a basis for flow forecasting wherever possible. Rainfall-runoff models are used to estimate flows on headwater and tributary subcatchments, and for estimation of lateral inflows. The potential for error in flood routing is far less than that in rainfall-runoff estimation so long as a high proportion of the inflow to a river reach can be estimated accurately in real time.

Flood routing models have traditionally been developed from theories of fluid flow. Such models have very high data requirements, and involve computationally difficult and slow numerical solutions. It is impractical to simplify such models to allow them to be calibrated and operated as part of a real-time flood forecasting system as required in the Severn and Trent basins. As an alternative, a simple conceptual model, which is quick to

Figure 11.1.   Severn–Trent water location map

execute and can be calibrated without excessive river surveys, is discussed
in this chapter. Efforts to overcome sources of error are presented, along
with examples of the use of the conceptual routing model as part of the
basin-wide flood forecasting package.

The flood warning service

Great Britain is a fairly small island and its longest river, the Severn, is only 280 km from source to tidal limit. Perhaps it is the preponderance of short, rapidly responding rivers which has led to British hydrologists' concentration on the rainfall-runoff process.

The Severn–Trent Water Authority has, over the decade since its formation, set about the development of a series of procedures for the region-wide forecasting of flooding. Much of the background to this effort was described by Jones (1980) and Manley *et al.* (1980). In this large region with diverse drainage patterns (Figure 11.1) flood warnings of expected flooding are provided not only of urban areas but also of low-lying agricultural land. Urban areas are, in the main, protected from frequently occurring floods. The standard of urban protection varies so that some towns might expect flooding to occur once in 5 years, whereas others with major defences would expect these to be overtopped less often than once in 100 years. Warnings are provided for flooding of agricultural land which might normally occur several times each year.

Warning time

Where a warning cannot be given in sufficient time to allow both for its dissemination to the population and for action to be taken, a service is not offered. Thus warnings are not provided for flash flooding in the headwaters of the Severn or the Trent and their many tributaries. However, where by alert observation and prompt forecasting a warning is possible at least 4 h before flooding starts, the service is offered. Developments of radar to improve short-term rainfall forecasting may enable the warning service to be extended to upstream reaches.

In contrast, towards the lower parts of the two river basins, warnings can often be given some 36 h or more in advance of flooding, this being the time taken for a flood originating in the headwaters to reach the lower reaches. In these locations, flooding may well last for days or weeks rather than hours, and it is naturally important to be in a position to advise on the likely duration of flooding, not solely on the maximum expected level of the flood waters.

Whilst a 'lead' time of 36 h is not typical, the majority of warnings are provided for places where there is a lag of 12 h or more between rainfall and flooding. This has two important major advantages:

(a) It allows time to collect and to analyse data from the network of tele-metering rainfall and river level gauges, and to use these data as inputs to mathematical models.

(b) It enables upstream river hydrographs to be used in the forecasting process in addition to rainfall information, so that flood routing models can be used in combination with rainfall-runoff models

FLOOD FORECASTING SYSTEM

No model, or suite of models, is sufficient on its own to enable a flood warning service to be offered. The model is just one link (albeit a very important one) in a chain, which also includes:

1. maintenance of field instruments and telemetry;
2. provision of competent staff to operate the service;
3. availability of weather information, both weather radar display, and precipitation forecasts;
4. collection of data from gauges and presentation to forecasting system;
5. data preparation routines;
6. network or system diagram linking the models together;
7. calibration of the models for each part of the basin;
8. interpretation by a duty officer of model forecasts in terms of areas/ properties expected to be flooded;
9. prompt dissemination of a warning to areas at risk.

Without proper attention being given to any one of these elements, the warning service will not function effectively. However, the organizational aspects will get little further mention here, since the emphasis in this chapter is on the forecasting system, and in particular the application of flow routing models in the production of flood forecasts.

A computer-based flow forecasting system has been developed to control the production of forecasts at numerous points throughout the two river basins using networks of rainfall-runoff and flow routing models. System diagrams were drawn up of the Severn and Trent basins which, to their respective tidal limits, include 66 and 49 tributary or lateral inflow catchments respectively, plus 48 flow routing reaches each. The forecasting system accepts and processes data collected by the computer telemetry scanners from a network of about 80 tipping bucket rain gauges and 130 river level gauges. Data are sorted, and quality control checks are carried out to identify any spurious data points, prior to the production of a large hourly data set for use with the forecasting models.

Because forecasts are required for more than 60 points in each basin, the duration associated with collection and preprocessing of the input data is comparatively long. It was therefore important to develop forecasting models which are themselves relatively simple, and quick and efficient to execute. Use of a telephone-based telemetry system means that a complete scan of all outstations takes almost 1 h. Allowing time for transfer of data from the telemetry computer to the forecasting computer, plus the 30 min or so taken by the models to produce the forecasts, means a minimum delay of some 90 min between collecting the first data and having the final forecast available. If forecasts are required more quickly for a small part of the basin then data collection and forecasting system operation can be geared accordingly, cutting the elapsed time to less than 1 h.

There are several fundamental considerations which enable forecasts to be produced efficiently by this type of system:

(a) Manual intervention in the system must be kept to a minimum. Data quality control, infill of missing data and entry of rainfall forecasts are the only elements requiring manual input to the system.
(b) All situations must be catered for in a robust way, so that the system will continue to operate in the event of the failure of any particular gauge or gauges.
(c) The models themselves must be efficient in their use of computer time, must be numerically stable, and must be a sufficient representation of the hydrological processes involved.
(d) Forecasts must be presented in a readily assimilated form, and be available at all times for transmission to operations personnel.

FORECASTING BY FLOW ROUTING

Attention here is focused on the role of the flow routing model in production of flood forecasts and warnings. The conceptual rainfall-runoff model developed for real-time use in the flow forecasting system was described by Bailey and Dobson (1981).

In rainfall-runoff models there are two essential components: the magnitude of the runoff (i.e. the proportion of the rainfall which produces the flood response) and the shape of the runoff hydrograph. Flow routing models are concerned solely with modification in the hydrograph shape, since the volume leaving the reach is the same as that entering it (in the absence of diversions, bypassing, evaporation or infiltration losses). Modelling of the flow routing process is therefore liable to much smaller errors than are inherent in rainfall-runoff models. In some cases downstream peak levels can be forecast using straightforward correlation graphs, but because of the number of major tributaries in both drainage networks, graphical methods are rarely totally satisfactory. Flow routing models extend the correlation principle to consider the whole flow hydrograph.

The theories relating to the flow of water in open channels were advanced in the nineteenth century by Poisson (1816), Boussinesq (1871) and culminated in the fundamental St Venant equations of conservation of mass and momentum (St Venant, 1871). Since there is no known analytical solution to these hyperbolic partial differential equations, their modification to provide a basis for description of flow in rivers has produced a wide range of approximate solutions. These typically involve linearization of the full St Venant equations (Lighthill and Whitham, 1955) or of a simplified version (Abbott, 1966) so that an analytical solution is feasible. Inevitably the solution is mathematically complex. In recent years this approach has developed to take advantage of computers, and has resulted in simplified linear representation of the St Venant equations solved by finite difference approximation. These iterative methods enable detailed analysis of wave hydraulics based on kinematic and diffusion models, and are used extensively for hydraulic design in river engineering (Samuels and Gray, 1982).

Other methods have been developed from different bases; the best known alternative is the Muskingum model (McCarthy, 1938). The Muskingum model is typical of a series of so-called hydrological models where flood routing is achieved by relating outflow to inflow and reach storage. Because they observe the principle of conservation of mass they can be shown to derive from the fundamental St Venant equations. The feature common to many of these hydrological methods is the need for *a priori* knowledge of current outflow in order to compute future reach outflow. Some achieve their objective using separate lag and route methods (Linsley *et al.*, 1949) but in others the two components are interlinked in the model formulation (Price, 1977).

Other flood routing models use mathematical operators to relate input and output data series. Impulse/response models, and associated stochastic methods, are not necessarily derived from any hydraulic or conceptual basis, and provide a flexible approach to the problem of flood routing (Powell and Cluckie, 1985).

Attempts to adapt many of the above models specifically for use in real-time flood forecasting have been made. In well-surveyed larger river basins, where flood routing models are well calibrated by reference to historic observations, there are some successful adaptations, notably for hydrological, kinematic and diffusion methods. Unfortunately the dependence of these models on iterative solutions by finite difference methods means they are susceptible to instability if the spatial or temporal interval used as the basis for linearization becomes too large. In many UK situations real-time forecasting is required on rivers which respond rapidly in their intensively used and populated upper reaches, and where lead times for forecasts are comparatively short.

To avoid these problems of instability, an empirical conceptual model, nicknamed DODO, was developed specifically for application in real time to a diverse range of river reaches in the Severn and Trent drainage basins.

DEVELOPMENT OF A REAL-TIME FLOW ROUTING MODEL: ESTIMATION OF REACH INFLOWS

Before flow routing can be undertaken, a data set has to be assembled of all inflows to the reach of river in question. Inflows to the reach may originate from one or more of:

(a) The measured or estimated outflow from an upstream routing reach.
(b) The measured or estimated outflow from a tributary reach or sub-catchment.
(c) The estimated lateral inflow to the reach.

Further, they might be inputs to the top of the reach, to the bottom of the reach, to some intermediate point or they might discharge uniformly along the length of the reach.

Because of the configuration of the flow routing model, it is convenient to consider inputs occurring only at either the top or bottom of a reach. Reach definition must take this simplification into account, such that reach boundaries would coincide with:

gauging stations;
tributary confluences;
points for which forecasts are required.

Inputs may be measured in real time at, for example, telemetry gauging stations on the main river or on tributaries. Many inputs will not be measured but must be estimated. Figure 11.2 illustrates a typical configuration, in which it is required to forecast flows at point F, using a flow routing model of the reach B–F. Inputs to the reach are:

(a) At the top of the reach: (i) output from the reach A–B, (ii) flow from the ungauged tributary C, (iii) flow from the gauged tributary D.
(b) At the bottom of the reach, flow from the gauged tributary E.
(c) At points along the reach, ungauged lateral inflows.

To give a sufficient lead time for the forecast at F it may be necessary to consider estimates of inflows to both the top and bottom of the reach not just up to the present time, but also input flows for some time ahead. Output from the reach A–B would have derived from gauged flows at A being applied to an upstream flow routing model to produce a simulated flow hydrograph at B. If a forecast hydrograph was available at A, the hydrograph at B could incorporate the lead time of the forecast at A plus the lag time between A and B. The flow hydrographs from the tributary C and from

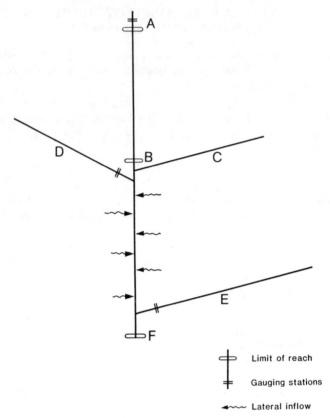

Figure 11.2. Inflows to a flow routing reach

lateral inflows are not measured, and must be estimated by applying the rainfall-runoff model. The lead time available would depend on the time of concentration inherent in the catchment's rainfall-runoff process, but might be extended by including rainfall forecasts for some hours ahead. In the forecasting system, flows from lateral inflow catchments are divided equally between the top and bottom of the reach.

Flows from tributary D are measured at a gauging station and, up to the present time, an observed hydrograph is available for input to reach B–F. This observed hydrograph can be extended into the future by applying the catchment rainfall profile (further extended if required by use of rainfall forecasts) to a rainfall-runoff model of this catchment. The summation of the flows from B, C and D provides a composite inflow hydrograph to the reach B–F, deriving from measured flows where available, from the output from earlier routing models and from rainfall-runoff models, and being extended into the future by using rainfall forecasts and the various natural lags in the hydrological system.

MODELLING IN-CHANNEL FLOWS

The model (DODO) requires only a sequence of inflows to each reach, and no *a priori* knowledge of reach outflows. It observes the principle of conservation of mass, and most importantly is solved quickly and directly using a small minicomputer. The search for a robust but simple real-time flood routing procedure began with a conceptual appraisal of the objectives for the model. When flood waves enter a channel reach at the upstream limit (usually a river gauging station) they change their nature as they move downstream to the outflow point. Comparison of the plots of river level at the upstream and downstream points shows (a) that there is a lag time as the floodwave moves downstream, and (b) that the shape of the floodwave is altered.

Normally, the wave is flatter and has a broader time base at the downstream point; it has attenuated. It was thought desirable to keep separate the model's treatment of lag and attenuation, thus simplifying model calibration and execution.

The lag time of floodwaves in a given river reach is not constant, but varies with the magnitude of the discharge in the reach. Small increases in flow in rivers during summer low flows move slowly through the river system. When baseflows are high and the increase in flow is larger, the rate of translation of flood waves is quicker. The same feature applies equally to controlled discharges from river regulation reservoirs.

The degree of departure from constant lag time in river reaches is conditional on the nature of the channel morphology, and the need for variable lag times was recognized in the variable parameter formulation of the Muskingum–Cunge method (Price, 1977). Analysis of the limited data available suggested that the greatest rate of change occurred between dry-weather discharges up to a point at about 75 per cent of the reach bankfull discharge. The increase in wave speed above this discharge was less marked. There is some evidence to suggest that in the early stages of an out-of-bank flood, wave speeds may reduce, although this may be due to the effects of floodplain storage, and will be discussed later.

The variable lag time is represented by a simple logarithmic relationship:

$$w = VC \cdot QI^{VE} \qquad (11.1)$$

Where:

QI = total inflow to the reach (m^3/s)

w = wave speed in m/s (maximum value VM),

VC, VE and VM = model parameters.

$$\text{Lag} = (L/w)/3600 \tag{11.2}$$

$$\text{Lag} = \text{lag time (h)}$$

$$L = \text{reach length (m)}$$

The function is truncated at a maximum wave speed *VM*; the point of truncation is determined during model calibration and is typically found to occur between 75 per cent and 100 per cent of bankfull discharge in the reach. Inflows to the upstream limit of a reach are lagged using this simple three-parameter function for all time steps in the forecasting interval, as the first stage of the routing procedure.

Once inflows have been lagged they are divided into in-bank and out-of-bank components. The reach bankfull discharge (QBF) is taken to represent the average channel conveyance between the upstream and downstream points. There are many cases where the bankfull discharge at river gauging stations is different from the average channel conveyance, since gauging sites are chosen to measure as much of the discharge as possible, and this may involve choice of sites which are not typical of the reach their measurements represent.

The in-bank component of the lagged inflow is attenuated using a two-parameter inflow–storage–outflow relationship:

$$QO_{(t)} = F1\,(QI_{(t\text{-Lag})}) + F2\,(SI_{(t)}) \tag{11.3}$$

Where:

$QO_{(t)}$	= in-bank reach *outflow* at time t;
$QI_{(t\text{-Lag})}$	= lagged in-bank reach *inflow* at time $(t\text{-Lag})$;
$SI_{(t)}$	= in-bank channel storage volume at time t;
$F1$ and $F2$	= model parameters with individual values between 0 and 1.

In order to preserve the principle of conservation of mass, channel storage is adjusted at each time step to account for the proportion of the inflow not discharged during the current calculation.

$$SI_{(t+1)} = (1-F2)SI_{(t)} + (1-F1)QI_{(t\text{-Lag})} \tag{11.4}$$

Where:

$SI_{(t+1)}$	= in-bank channel storage volume computed for the next time step.

Because the reach lag is determined by a separate function, the behaviour of the inflow/outflow and storage increment/decrement components of the routing function is very simple.

When channel inflow increases to the bankfull for the reach, the function for in-bank attenuation reaches a steady state, with constant channel storage (SI) and an outflow equal to the fixed bankfull discharge. Surplus volumes above bankfull are considered separately.

MODELLING OUT-OF-BANK FLOWS

Many rivers have flood banks, either natural levees or man-made constructions. When the flood banks are designed to afford a high degree of flood protection, usually to protect urban areas, then they effectively increase the reach bankfull discharge, and the model will operate as described above until design conditions are exceeded. When banks are overtopped—whether this occurs frequently over natural or low artificial flood banks, or rarely in urban flood defence schemes—a second routing procedure provides for attenuation of the out-of-bank components of flow.

The out-of-bank routing procedure has two main components: the *first* deals with static storage of water on floodplains, and the *second* with water flowing on the floodplain. If the floodplain slopes towards the channel and there are no floodbanks or natural levees then the first component of the out-of-bank routing procedure is not required.

There are, however, many reaches where the initial component of the out-of-bank flow flows over banks into areas of static floodplain storage. If the areas of static storage are designed for flood alleviation then evacuation of the flooded areas may be achieved through flapped outfalls which operate once flows have returned to within-bank. Alternatively, the static storage may return to the channel through a low point in floodbanks.

All inflow above bankfull is considered to enter the reservoir of static floodplain storage until its capacity (SOB) is filled. The capacity of the static floodplain storage area is either determined during model calibration, or can be estimated by field survey. The importance of static floodplain storage is often greatest in confluences between major rivers, where the floodplain has evolved to accommodate coincident floods from each tributary, and therefore offers a comparatively high storage volume. When floods from the tributaries do not coincide in time, then the attenuation provided by static storage can modify two separate flood peaks to such an extent that levels below the confluence are held at bankfull for a long period. An excellent example of this phenomenon is found at the confluence of the rivers Severn and Vyrnwy upstream of Shrewsbury (see Figure 11.3).

Controlled washland storage areas behave in much the same way, with outflows limited to bankfull until the available storage is filled. In many

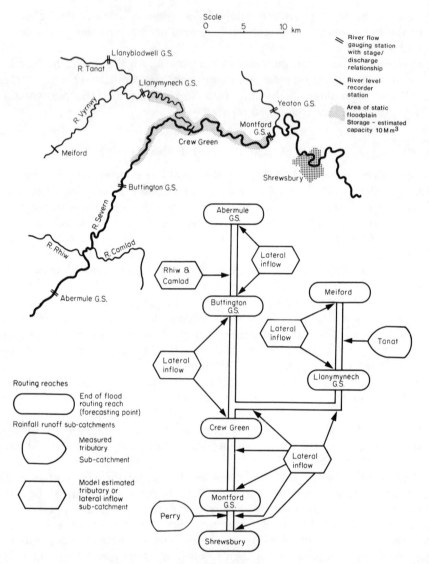

Figure 11.3. Severn–Vyrnwy confluence to Shrewsbury location map and model representation

cases the volume available for storage is known when such areas are designed as part of a local flood alleviation scheme.

Once excess flows above bankfull have filled the volume of available storage, then provided reach inflow still exceeds bankfull, the second component of the out-of-bank routing procedure will attenuate the dynamic out-of-bank flows. The method used for attenuation of out-of-bank flows is identical to the inflow–storage–outflow procedure described for in-bank routing, but with different coefficients and using a second storage account for dynamic (i.e. flowing) floodplain storage:

$$QO^*_{(t)} = F3\,(QI^*_{(t\text{-Lag})}) + F4\,(SO_{(t)}) \tag{11.5}$$

Where:

$QO^*_{(t)}$ = dynamic floodplain *outflow* at time t;
$QI^*_{(t\text{-Lag})}$ = dynamic floodplain *inflow* at time (t-Lag);
$SO_{(t)}$ = dynamic floodplain storage volume at time t;
$F3$ and $F4$ = model parameters with individual values between 0 and 1.

The account for dynamic floodplain storage is updated at each time step as for the inbank routing:

$$SO_{(t+1)} = (1-F4)SO_{(t)} + (1-F3)QI^*_{(t\text{-Lag})} \tag{11.6}$$

Where:

$SO_{(t+1)}$ = dynamic floodplain storage volume computed for the next time step.

Once the outflow peak has passed, and the inflow reduces below the rate of outflow, then the emphasis shifts to the reduction in dynamic storage on the floodplain. When outflow has reduced to the reach bankfull discharge, then the water held in the static storage reservoir on the floodplain can begin to empty freely. Before this time, however, a limited amount of drainage from static storage is permitted, to simulate the performance of outflow controls during the falling stage of a flood.

The degree of constraint on outflow from static storage is determined in real time by reference to the magnitude of the peak outflow. The constraint is progressively reduced to zero at the point where static storage outflow is unimpeded by in-channel river levels. In the case of submerged flapped outfalls this point may not be reached until flows in the channel have receded to a point below reach bankfull discharge, and the model is able to simulate this effect by reference to a return bankfull discharge (QRF). This is chosen to be equal to the reach bankfull discharge or some suitable lower flow. The unimpeded flow of water from static storage back into the channels is achieved using a simple logarithmic function:

$$QR_{(t)} = CRS(SS_{(t)})^{ERS} \tag{11.7}$$

Where:

$QR_{(t)}$ = return flow from static storage at time t;

$SS_{(t)}$ = current quantity in static storage at time t (maximum
                value SOB); *CRS*, *ERS*, and *SOB* model parameters.

MODEL CALIBRATION AND APPLICATIONS

The model is therefore a simple conceptual representation of flood wave modification. It achieves its objective by dealing separately, so far as is possible, with the lagging and attenuation procedures. There are twelve parameters controlling the routing and these are calibrated in groups of three or four, where each group relates to a particular procedure in the model.

Some model parameters can be estimated easily by reference to historic floods, e.g. reach bankfull discharge and return bankfull discharge, and maximum wave speed. The other parameters are derived in calibration of the model using data from historic floods. Investigation is moving towards assessment of parameter sensitivity and the relationship between parameter choice and model stability. Further work is being directed to the development of methods to relate optimum parameter values for the model to physical and morphological characteristics of the river reach.

Results from model calibration and operational implementation show that the model is flexible enough to cope with many different types of routing reach from fast-flowing steep streams to mature lowland river reaches. It is especially valuable in complex confluence areas, and where floodplain storage is a potent factor in flood wave modification (see Figure 11.4).

The model remains stable using a 1 h timestep for all reaches in the Severn–Trent areas, apart from occasional instabilities resulting from large increases in reservoir discharge occurring in a single timestep, or from rapid oscillations in input data series caused by missing or inaccurate data. Both of these features affect other routing models, often producing worse instability. The effects of input data errors or missing data points are minimized by the application, in real time, of a simple error correction/updating model.

SUMMARY AND PROPOSALS FOR ENHANCEMENT

The rainfall-runoff model described earlier, and the flood routing procedure described here, together with a simple error tracking and updating algorithm, form the nucleus of the Severn–Trent flow forecasting system. The system has been operational since 1982. The current system uses telephone-based telemetry data collection by Plessey minicomputers and Delta Technical Services TG4000 processor. The flow forecasting suite is mounted on IBM Series I minicomputers, and is written in FORTRAN.

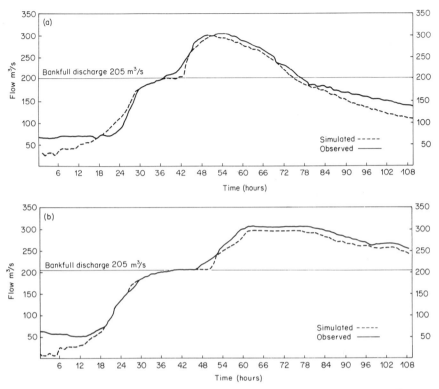

Figure 11.4. Examples of routing model performance for River Severn at Shrewsbury

Implementation of the computer-based systems highlighted the practical problems which need to be resolved for successful real-time forecasting. A common cause of poor model performance is unreliable or inefficient collection of data in real time. Data management routines, and logistic considerations associated with efficient model running, need careful design and implementation. By contrast, in hydrological simulation these potential difficulties are overcome during data preparation, and even if the same mathematical formulation is used, the particular problems of operation in real time are not met.

Model calibration is a considerable task, and is one which needs periodic review. River maintenance schemes, or flood alleviation measures, can significantly alter the behaviour of a tributary or river reach. The importance of good hydrometric measurement is paramount in achieving a high-quality data input. The present flow forecasting system is considered to make the best use of the data currently available in real time, especially in the application of the flow routing models.

Improvements to the flow forecasting system are concentrating on stream-lining the data acquisition and preparation into a satisfactory data sequence for modelling. The system requires a continuous sequence of hourly data to update state variables in the rainfall-runoff model and the flow routing model. This ensures that information on antecedent catchment conditions is available before the start of each event without the need for initialization. The selection of this type of system affects the choice of enhancements.

Radar rainfall measurements, although generally reliable, are prone to total breaks in data availability. Because failure at a radar site results in loss of good-quality areal rainfall measurements over a large area, it is not possible to substitute radar measurements in place of a network of telemetry rain gauges. Rather it is envisaged that radar rainfall measurements over predefined subcatchments will be used to enhance the data estimated from the network of rain gauges. Relative intensities within rainfall systems are clearly depicted by radar, and even if the actual intensity measurement is in error, the use of 'ground truth' from the network of raingauges enables the two sources of rainfall measurement to be used conjunctively to improve estimates of both magnitude and spatial variations in rainfall.

Telemetry outstation technology has improved considerably, with instruments now able to log rainfall and also river levels data at preset intervals on site.

Interrogation of these intelligent data loggers prior to forecasting enables the required hourly data to be retrieved in one poll of the network, rather than the present system which polls at frequent intervals to retrieve the current status only. Incorporation of these new telemetry devices will allow a faster interrogation of a better-quality data sequence. This will significantly reduce the constraints and difficulties in data collection mentioned earlier.

River flow measurement at sites where conventional level and stage/discharge methods are unattainable is now available using ultrasonic and electromagnetic gauging stations. These developments enable flow data to be available at difficult sites, often the sites where forecasting models most need it.

Incorporation of radar rainfall measurements and improved telemetry will enable forecasts to be produced more rapidly for locations further upstream on the river network. It will also allow improvements in the quality of model forecasts by providing a logged hourly dataset. Response earlier in an event, when rain is still falling, may be made more reliable if radar rainfall measurement allows accurate quantitative precipitation forecasts over predefined areas (see Sargent, Chapter 3 and Browning, Chapter 16, in this volume).

ACKNOWLEDGEMENT

The authors acknowledge permission to publish granted by the Regional Manager, Rivers and Resources, Severn–Trent Water Authority. The views expressed are those of the authors.

REFERENCES

Abbott, M. B. (1966). *An Introduction to the Method of Characteristics.* New York: American Elsevier.
Bailey, R. A., and Dobson, C. (1981). Forecasting for floods in the Severn Catchment. *J. Inst. Water Engrs Sci.,* **35**(2), 168–78.
Boussinesq, J. (1871). Theory of the liquid intumescence, called a solitary wave or a wave of translation, propagated in a channel of rectangular cross section. *Comptes Rendus Acad. Sci., Paris,* **72**, 755–59.
Jones, H. H. (1980). An overview of hydrological forecasting in a multi-functional Water Authority. *Proc. Oxford Symposium,* IAHS publ. no. 129, pp. 195–202.
Lighthill, M. J., and Whitham, G. B. (1955). On kinematic floods, I: Flood movements in long rivers. *Proc. Roy. Soc. Lond.,* **A229**, 281–316.
Linsley, R. K., Kohler, M. A., and Paulhus, J. L. H. (1949). *Applied Hydrology,* New York: McGraw-Hill, pp. 502–30.
Manley, R. E., Douglas, J. R., and Pirt, J. (1980). Conceptual models in a flow forecasting system. *Proc. Oxford Symposium,* IAHS publ. no. 129, pp. 467–78.
McCarthy, G. T. (1938). The unit hydrograph and flood routing. Paper presented at Conf. of the North Atlantic Div. of US Corps of Engineers, New London, Connecticut.
Poisson, S. D. (1816). Memoir on the theory of waves. *Memoirs,* Vol. 1. *Acad. Sci., Paris,* pp. 71–186.
Powell, S. M., and Cluckie, I. D. (1985). Mathematical hydraulic models for the real time analysis of floods. BHRA 2nd International Conference on the Hydraulics of Floods and Flood Control.
Price, R. K. (1977). FLOUT—a river catchment flood model. Hydraulics Research Station. *Report IT168.*
Saint-Venant, B. de (1871). Theory of unsteady water flow, with application to river floods and to propagation of tides in river channels. *Comptes Rendus Acad. Sci., Paris,* **73**, 148–154, 237–240. (Translated into English by US Corps of Engineers, No. 49-g, Waterways Experiment Station, Vicksburg, Mississippi, 1949.)
Samuels, P. G., and Gray, M. P. (1982). The FLUCOMP river model package—an engineer's guide. Hydraulics Research Station Report Ex. 999.

Weather Radar and Flood Forecasting
Edited by V.K. Collinge and C. Kirby
© 1987 John Wiley & Sons Ltd.

CHAPTER 12

Real-time Rainfall-Runoff Models and Use of Weather Radar Information

I. D. CLUCKIE AND M. D. OWENS

INTRODUCTION

The North-West Radar Project was established in 1977 with a major objective of incorporating real-time weather radar data into an operational flood forecasting system. One aspect of the project, namely the development of rainfall-runoff models for flow forecasting at selected sites within the North West Water Authority area, has been considered by Knowles in Chapter 10. It is discussed further in this chapter, which describes a mathematical study of event-based flood forecasting models using a combination of weather radar-derived rainfall estimates and conventional telemetry. Three catchments with significant flood risk potential have been studied; the case study presented here focuses attention on one particular example, the Blackford Bridge gauging station in the Irwell catchment.

Transfer function models have been developed to provide real-time flow forecasts for catchments in the area using radar-derived precipitation data. These models use past flow and total rainfall to forecast future flow. The fact that this type of model is economic in resource utilization is an important feature when modelling large catchments in an on-line situation. The flow forecasts are provided for a lead time of 6 h ahead, which is regarded as the optimum for flood warning procedures currently employed in the United Kingdom.

Model structure, parameter estimation, basic sampling interval and the effect of degraded data are all discussed in relation to the practical use of such models in event-based flood forecasting systems. Attention is given to the incorporation of future rainfall scenarios and in particular short-term precipitation forecasts produced by the FRONTIERS system developed by the Meteorological Office.

THE RAINFALL-RUNOFF MODEL

The event-based rainfall/runoff model developed is a simple linear transfer function of the following form:

$$y_t = a_1 y_{t-1} + a_2 y_{t-2} + \ldots + a_p y_{t-p} + b_1 u_{t-1} + b_2 u_{t-2} + \ldots + b_q u_{t-q}$$
$$(12.1)$$

Where:

y_t	= the flow at time t;
u_t	= the total rainfall at time t;
a_i and b_i	= model parameters

The model is simple both to apply and understand by the user. Unlike a physically based model it has minimum data requirements which are easily obtained in real time via a telemetry system. The transfer function model is a numerical approximation to the traditional unit hydrograph (Sherman, 1932), and is subject to the same primary assumptions of system linearity and time invariance. The on-line input of previous flow measurements in addition to past rainfall values results in a parsimonious model structure which is easily initialized at the beginning of an event and naturally provides the model with a self-correcting feedback which is absent in the traditional unit hydrograph approach which possesses a closed-loop structure. The impulse response of the model, which is defined as the flow due to an input of unit rainfall, is analogous to the unit hydrograph. The impulse response is an important characteristic used to identify an appropriate hydrological model for producing multi-step forecasts.

The unit hydrograph theory is based upon the assumed linear relationship between direct runoff and effective rainfall, and ensures a volume balance. Many methods of estimating effective rainfall have been proposed (e.g. NERC, 1975). The difficulty of defining effective rainfall, particularly in a real-time situation, has led to the development of a model utilizing an input of total rainfall. The baseflow component is assumed to be constant at a level equal to the lowest flow prior to the event. A model derived using total rainfall thus has a steady-state gain which preserves the percentage runoff of the particular event sequence used for parameter estimation.

The response of a catchment is time-variant and is not strictly linear. Forecasting errors are caused by fundamental data errors and also the inadequacy of the chosen model to fully synthesize the response of the hydrological system to the complete range of possible inputs. The errors can produce a combination of timing and magnitude problems which are difficult to identify and separate. A model with a steady-state gain representing a vastly different percentage runoff from that of a specific event will produce poor forecasts with predominantly magnitude errors. An approach to cope with both timing and magnitude errors is to update the model parameters in real time, thus changing the impulse response shape in accordance with the desire to minimize the forecast errors, as often practised in control systems. An investigation by Harpin (1982) revealed that, in the real-time hydrological environment where the rainfall-runoff model is required to be adequate on the rising limb of the flood event when very few data points have become available, real-time updating of parameters ws not practical. This is primarily due to the slow convergence rates of such recursive parameter estimation schemes in an environment where the information content of the incoming data is so minimal.

An alternative simple approach, aimed at updating the percentage run-off represented by the model, was introduced by Cluckie and Smith (1980). A real-time correction factor, delta, scales the rainfall parameters of the model to match the model's steady-state gain with the event percentage runoff in the following manner:

$$y_t = a_1 y_{t-1} + a_2 y_{t-2} + \ldots + a_p y_{t-p} + \Delta b_1 u_{t-1} + \Delta b_2 u_{t-2} + \ldots + \Delta b_q u_{t-q} \quad (12.2)$$

The percentage runoff is only accurately determined at the end of an event, so the one-step-ahead forecast error is used to update delta. In this manner all the forecast error is attributed to the incorrect estimation of the percentage run-off. Ideally, delta must be able to adapt quickly to new information on the rising limb of an event, but must also be sufficiently smoothed to dampen any erratic variations caused by data errors.

Owens (1986) has used a method of updating delta of the following form:

$$\Delta_t = \mu \Delta_{t-1} + (1-\mu) \left(\frac{y_t - (a_1 y_{t-1} + a_2 y_{t-2} + \ldots + a_p y_{t-p})}{b_1 u_{t-1} + b_2 u_{t-2} + \ldots + b_q u_{t-q}} \right) \quad (12.3)$$

Where:

$0 \leqslant \mu \leqslant 1$ is a smoothing factor.

An example showing the advantage of using delta within the forecasting procedure has been included in this case study.

WEATHER RADAR FOR HYDROLOGICAL USE

The spatial information on rainfall over a large area as provided by the Meteorological Office radar network is invaluable for the recognition of storm development and the consequent production of rainfall forecasts. These can then be used as essential inputs into hydrological models. The ability to replay storm sequences enables the duty hydrologist to assess the local flood situation within an overall temporal and spatial context, thus aiding the decision-making process within the flood warning system. When quantitative radar estimates are used in hydrological modelling the reliability of the radar is crucial, since failure affects the whole database and infilling is difficult or impossible, unlike the failure of individual raingauges within an extensive network. Collier (1985) has indicated that the availability of Hameldon Hill radar is greater than 90 per cent, including periods of scheduled preventive maintenance. In addition, the measurement of snowfall, which is difficult by raingauge, can be reasonably accurately estimated using radar (Collier and Larke, 1978).

A detailed study of the accuracy of radar estimates, in an operational context, was carried out within the North-West Radar Project (Collier, 1985 and Chapter 6, this volume). Accuracy can only be determined for a point estimate over a raingauge, assuming that the rain gauge itself gives an accurate estimate of true rainfall. This type of measurement of accuracy of the radar may not be the most important practical criterion on which to assess the suitability of radar data for hydrological forecasting. However, the nature of the radar data and errors encountered in the rainfall estimates must be recognized, so that hydrological forecasting procedures can be developed to cope with anticipated problems characterized by radar estimation errors.

DATA AVAILABLE

The North West Water Authority flood forecasting system (Noonan, Chapter 8, this volume) receives data every 15 min from a telemetry network of raingauges and river gauges, as well as receiving precipitation estimates from the unmanned weather radar at Hameldon Hill, as indicated in Figure 12.1. The radar rainfall intensity information is derived at a resolution of 2 km to a range of 75 km and a resolution of 5 km to a range of 210 km. Fifteen-minute rainfall totals, derived from the 2 km information, are available for the NWWA hydrologically defined catchments. These data are transmitted at 208 levels (eight-bit data). Rainfall intensity information is transmitted at eight levels (three-bit data) for display on a colour graphics device.

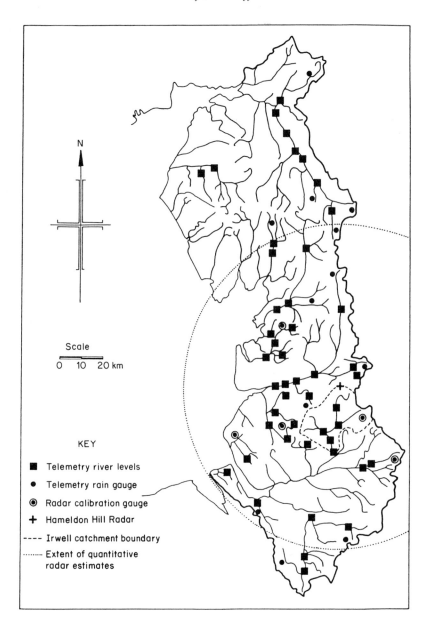

Figure 12.1   North West Water Authority radar and telemetry network

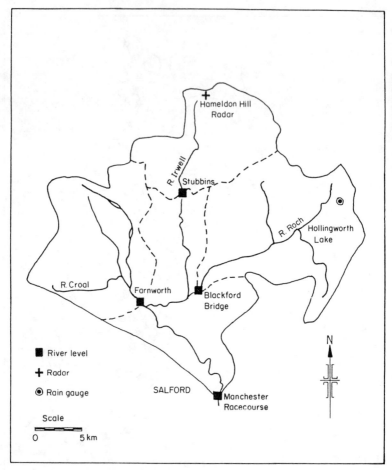

Figure 12.2.   The Irwell Catchment

CASE STUDY—THE IRWELL CATCHMENT

The Irwell Catchment (Figure 12.2), which is situated in the Southern Pennines, is predominantly elevated moorland with significant urban areas. The confluence of the three main tributaries occurs upstream of Manchester. The upper reaches of all tributaries experience minor flooding problems in the narrow plains adjacent to the river channel. The main flood risk occurs at Salford, where the floodplain extends over a widespread area including industrial and residential areas. It is therefore beneficial to forecast levels at Manchester Racecourse. Hameldon Hill radar is at the northern extent of

Table 12.1. Gauging stations in the Irwell Catchment

River	Gauging station	Area (km)	Land use
Upper Irwell	Stubbins	105	moorland
Roch	Blackford Bridge	186	moorland, urban
Croal	Farnworth	145	moorland, urban
Irwell	Manchester Racecourse	557	moorland, urban

the catchment, and Hollingworth Lake calibration gauge lies within the Roch subcatchment. Details of the telemetry river gauges are given in Table 12.1.

PARAMETER ESTIMATION

The recursive least-squares algorithm, originally derived by Plackett (1950), was used to estimate the model parameters from a sequence of events. Harpin (1982) investigated several parameter estimation algorithms and suggested that a least-squares estimator was adequate for simple rainfall-runoff transfer function models, despite producing biased estimates when the model errors were serially correlated. Ede and Cluckie (1985) concluded that a model estimated using recursive least-squares can forecast as well as, or occasionally even better than, a model derived using a more sophisticated instrumental variables algorithm which produces unbiased parameter estimates.

The sequence of events chosen for parameter estimation must represent a range of peak flows and catchment conditions. It is important that the resultant average model reflects the catchment response over the fullest possible range of conditions. The main constraint imposed on the selection of hydrological events was that the radar data calibrated using the domain-based orographic procedure were not available prior to January 1982. The event sequence, consisting of six events for Blackford Bridge, is shown in Figure 12.3. The impulse response of the resultant hourly model is shown in Figure 12.4. The time to peak is 4 h and the shape is the familiar form associated with a typical unit hydrograph. Although the derived models from individual events are not suitable for forecasting, they can provide a valuable indication of the variability of the catchment response. The impulse responses of individual events as given in Figure 12.4 clearly show variability in magnitude, timing and recession shape.

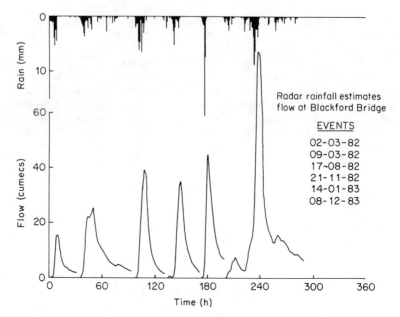

Figure 12.3.   Event sequence for parameter estimation

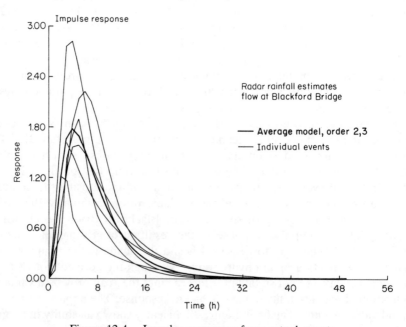

Figure 12.4.   Impulse responses from actual events

MODEL ORDER

The order of a model is the number of parameters (p and q in equation 12.1). The optimum model will have sufficient parameters to adequately represent the catchment response without being over-parameterized. The best method of obtaining the optimum model order for transfer function rainfall/runoff models derived using the recursive least-squares estimator has been investigated by Owens (1986), and is not discussed in this chapter.

MODEL SAMPLING INTERVAL

The rainfall and runoff data are available every 15 min. However, increasing the model discretization to 1 h for the Blackford Bridge model did not cause any significant difference to the impulse response, as indicated in Figure 12.5. The optimum hourly model is also more parsimonious. Increasing the model interval to 2 h did not result in a more parsimonious model, and began to cause a deterioration in the definition of the impulse response, indicating a loss of information. Generally the optimum model interval is related to the catchment response characteristics. Several authors, including Powell and Cluckie (1985) have suggested empirical rules for obtaining the optimum model interval, though it is possible to use an hourly model to produce forecasts for 15 min intervals by simply increasing the rate at which the input data are consolidated into 'hourly' blocks.

FORECASTING AND THE USE OF DELTA

A sequence of flow forecasts, both with and without the use of delta, is given in Figure 12.6. Rainfall spatial distributions at corresponding times are provided in Figure 12.7. These indicate the highly dynamic nature of this particular storm which moved in a generally south-westerly direction whilst undergoing considerable internal growth and decay. The model used for forecasting is the hourly model whose impulse is shown in Figure 12.4. The steady-state gain of this model corresponds to approximately 25 per cent runoff. The 14 March 1982 event had an actual overall percentage runoff of 77 per cent, although in a real-time situation this would not be known until the end of the event. The high percentage runoff was caused by the catchment being in a saturated condition, due to the occurrence of two smaller events on 2 and 9 March 1982.

The flow is forecast for up to 6 h ahead, which is the optimum lead time required by NWWA. Perfect foresight of rainfall is assumed, which allows the assessment of the performance of the delta factor. Without the use of delta the flow is persistently underestimated throughout the storm, despite the use of perfect foresight of future rainfall. Delta was initialized at unity

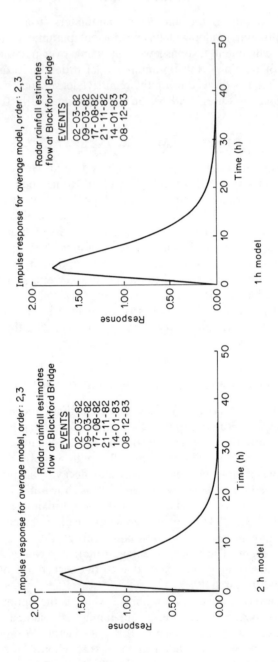

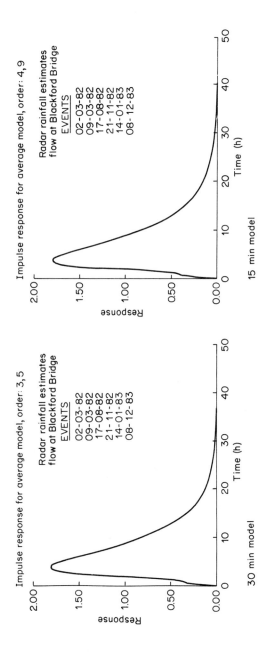

Figure 12.5. Impulse responses for models with varying discretization

182

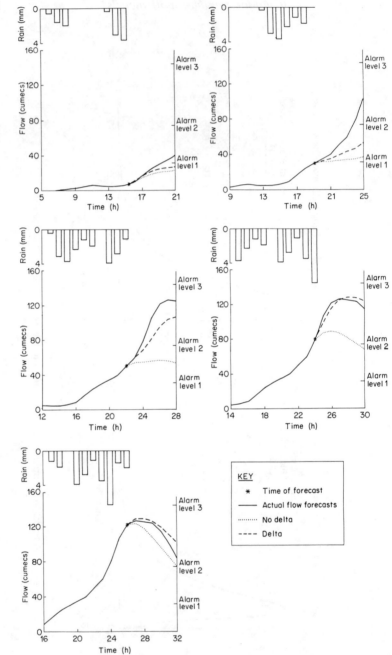

Figure 12.6. Forecast sequence for flow at Blackford Bridge, 14 March 1982, showing effect of using delta updating

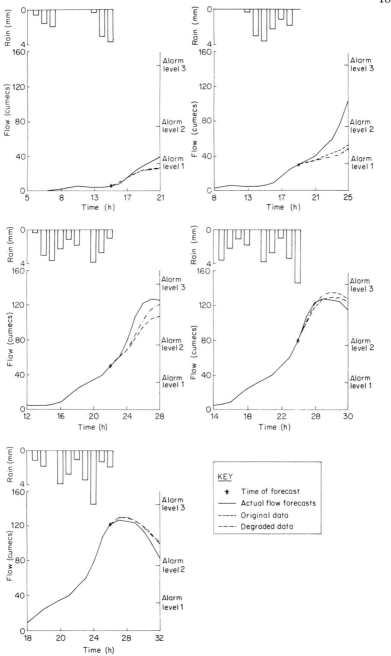

Figure 12.7. Forecast sequence for flow at Blackford Bridge, 14 March 1982, showing effect of using degraded data

and allowed to update throughout the storm. As the value of delta increases on the rising limb, so the forecasts become progressively better. This procedure for updating has been investigated in some detail and will be reported upon in due course. The importance of the need to calibrate the radar data in real time with reference to the ground raingauge measurements and the additional need to calibrate the rainfall-runoff model in real time was discussed in further detail in Collier and Cluckie (1985). The calibration of radar-derived quantitative rainfall measurements should not be seen in isolation from the hydrological use to which those measurements will eventually be put.

DEGRADED DATA

As mentioned previously, the subcatchment data is transmitted to the water authority after being quantized at 208 levels. The radar composite 5 km data are transmitted at eight intensity levels. There are distinct savings to be made in transmitting low-resolution data in real time, so a brief study has been undertaken to investigate the consequence of degrading the original subcatchment data to a resolution of eight intensity levels.

The rainfall within each range was assigned the value of:

(a) the arithmetic value, and
(b) the logarithmic average of the range.

The impulse responses derived from the original data and the degraded data were very similar in overall shape, with apparently only minor magnitude differences occurring. When the impulses were scaled the basic shape of each was almost identical. The scaling differences can be accommodated by the steady-state gains of the respective models when applied in real time.

The forecasting sequences produced for 14 March 1982 using the original and degraded data are shown in Figure 12.7. Delta compensates for any rainfall scaling, and the use of degraded data does not seem to result in significantly poorer flow forecasts. This applies in the case of both arithmetic and logarithmic level slicing procedures.

FRONTIERS DATA

The FRONTIERS system (Forecasting Rain Optimized using New Techniques of Interactively Enhanced Radar and Satellite), which is still under development at the Meteorological Office, is designed to produce quantitative rainfall forecasts for up to 6 h ahead on a 30 min cycle at a spatial resolution of 20 km. The system is described in detail by Sargent (Chapter

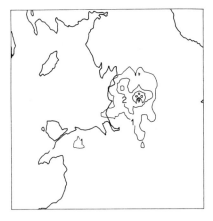

(a) Hameldon Hill actual 12.00
    (5 km grid)

Contour	Rainfall
1	2.5 mm/h
2	5.0 mm/h
3	7.5 mm/h
4	10.0 mm/h

(b) FRONTIERS actual 12.00
    (20 km grid)

Contour	Rainfall
1	2.5 mm/h
2	5.0 mm/h
3	7.5 mm/h
4	10.0 mm/h

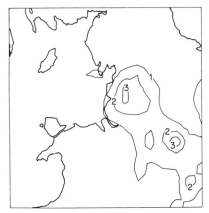

(c) FRONTIERS actual 12.00
    (20 km grid)

Contour	Rainfall
1	1.0 mm/h
2	2.0 mm/h
3	3.0 mm/h
4	4.0 mm/h

Figure 12.8.   Rainfall distribution, 29 January 1985

3, this volume) and further possible developments of the system by Browning (Chapter 16, this volume).

Although the system was still going through pre-operational trials, a single hydrological event has been analysed using the FRONTIERS data for illustration. It is not possible to draw any general conclusions on such a limited study, although it provides a tentative indication of the great potential of this product for hydrological forecasting.

The remote site Hameldon Hill 5 km resolution rainfall spatial distribution at 1200 hours on 29 January 1985 is shown in Figure 12.8(a). The corresponding FRONTIERS actuals are given in Figure 12.8(b) and (c). The two distributions differ for three main reasons:

(a) The FRONTIERS data have a spatial resolution of 20 km, so the average rainfall intensity over a 20 km square will be less than the peak 5 km resolution intensity. Therefore the computer-based contouring routine delineates a reduced area. By changing the contour slicing, as in Figure 12.8(c), a spatially more significant rainfall pattern is revealed.
(b) The FRONTIERS actuals (Figure 12.8(c)) which are derived using the composite radars in the UK network, are showing an area of rainfall which is not detected at range by the Hameldon Hill radar (Figure 12.8(a)),
(c) The FRONTIERS actuals will also differ from the remote site distribution because FRONTIERS estimates undergo subjective corrections to improve the quality of the product (i.e. interactively enhanced).

Sequences of FRONTIERS forecasts for 29 January 1985 at 1100 hours are shown in Figure 12.9(a). The FRONTIERS actuals show a large cell of rainfall which was forecast to move in an easterly direction. The two hour forecast can be directly compared to the actuals at 1300 in Figure 12.9(b).

At present the future rainfall scenarios available to the hydrologist are:

(a) no more rainfall, and
(b) past average rainfall.

These, together with the FRONTIERS forecasts and perfect foresight are shown in the forecast sequence provided in Figure 12.10. The perfect foresight scenario is included to give an indication of the performance of the model and is not available in a real-time situation. The no-more rainfall option gives a lower limit to the flow forecast and produces good forecasts after the rainfall event is complete. This highlights the essential requirement for rainfall forecasts, to allow the preparation of meaningful flow forecasts for up to 6 h ahead for a catchment with a response time of only 4 h. The past average rainfall options produce fairly good forecasts in situations of

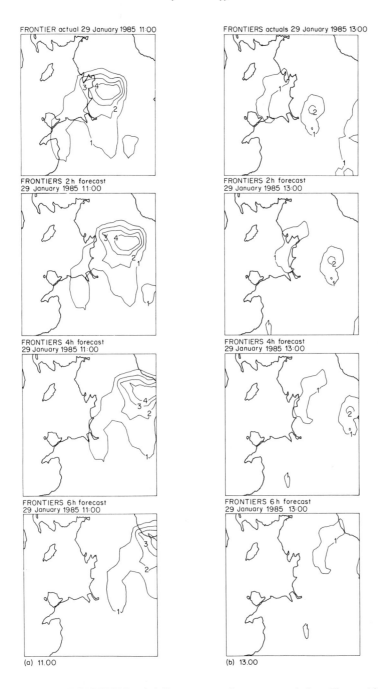

Figure 12.9. FRONTIERS rainfall sequences (contours scaled as Figure 12.8)

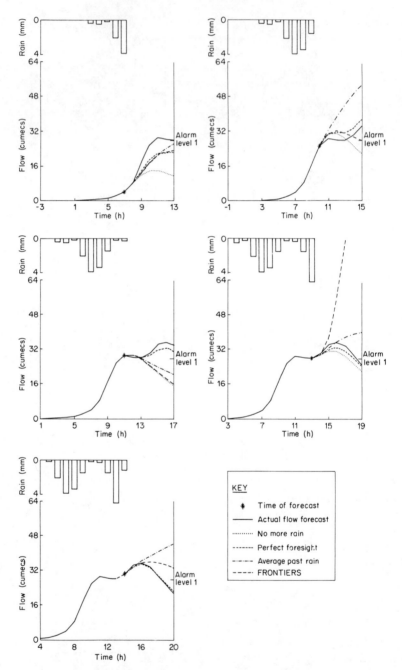

Figure 12.10. Forecast sequence for flow at Blackford Bridge, 29 January 1985, showing various future rainfall scenarios

constant rainfall, but will underestimate during significant storm development and overestimate during significant storm decay.

In this event the FRONTIERS data generally provide a helpful forecast and could be useful to the hydrologist. However, the overestimation for the forecast made 12 h into the event signals that this is rather too soon to make any firm conclusions, and more events must be analysed before making general recommendations of a hydrological nature. It is likely that it will take a further 2–3 years to fully appreciate the benefits of such a product to the hydrological community, though the initial indications are encouraging.

CONCLUDING COMMENTS

The potential for the use of quantitative precipitation estimates from a network of weather radars is clearly established. In particular the application of simple transfer function models to flow forecasting is shown to have considerable utility in the real-time environment. The complexity of the parameter estimation exercise is simplified by employing static rather than dynamic models with updating of the model's steady-state gain during the course of the event by using an error feedback technique.

The effect of using data with a reduced quantization level (referred to as degraded data) has been shown to have little effect on the events tested during the course of this work. The hydrological application of the FRONTIERS data product has been illustrated with the use of a case study. The potential when this type of data is used in a real-time hydrological modelling context is discussed in a preliminary manner.

ACKNOWLEDGEMENTS

The radar and rain gauge data used for the study were collected during the North-West Radar Project, a joint programme supported by the Meteorological Office; the North West Water Authority; the Water Research Centre; the Central Water Planning Unit (now defunct) and the Ministry of Agriculture, Fisheries and Food. The Science and Engineering Research Council, the North-West Radar Consortium and the North West Water Authority provided additional financial support.

REFERENCES

Cluckie, I. D., and Smith, F. J. B. (1980). Flood forecasting project for Wessex Water Authority (Avon and Dorset Division), Final Report, Vol. 1.
Collier, C. G. (1985). Rainfall estimates made by radar used in operational rainfall and flood forecasting system. Part I: The accuracy of the radar estimates calibrated in real-time using raingauge data. *J. Hydrol.* (In press).
Collier, C. G., and Cluckie, I. D. (1985). A hydrological study of the real-time

calibration of radar derived rainfall data. *Proc. Int. Symp. Advances in Water Engineering*. University of Birmingham, 15–19 July. London: Elsevier.

Collier, C. G., and Larke, P. R. (1978). A case study of the measurement of snowfall by radar; an assessment of accuracy. *Quart. J. R. Met. Soc.*, **104**, 615–21.

Ede, P. F., and Cluckie, I. D. (1985). End-point use as a criterion for model assessment. *7th IFAC/IFORS Symposium on Identification and System Parameter Estimation*, York, UK, pp. 475–80.

Harpin, R. (1982). Real-time flood routing with particular emphasis on Linear methods and recursive estimation techniques. Ph.D. thesis (unpublished), University of Birmingham.

NERC. (1975). Flood Studies Report, Vol. 1.

Owens, M. D. (1986). Real-time flow forecasting using Weather Radar data. Ph.D. thesis (unpublished), University of Birmingham.

Plackett, R. L. (1950). Some theorems in least squares. *Biometrica* **37**, 149–57.

Powell, S. M., and Cluckie, I. D. (1985). On the sampling interval of discrete transfer function models of the rainfall–runoff process. *Proc. 7th IFAC/IFORS Symposium on Identification and System Parameter Estimation*, York, UK, pp. 1119–1124.

Sherman, L. K. (1932). Streamflow from rainfall by the Unit-graph method. *Eng. Newsrecord*, **108**, 501–5.

Weather Radar and Flood Forecasting
Edited by V.K. Collinge and C. Kirby
© 1987 John Wiley & Sons Ltd.

CHAPTER 13

Flood Forecasting Based on Rainfall Radar Measurement and Stochastic Rainfall Forecasting in the Federal Republic of Germany

GERT A. SCHULTZ

INTRODUCTION

There are two main objectives for which flood forecasts are computed and issued. These are for flood warning to reduce loss of life and damage to goods and property due to floods, and operation of flood protection reservoirs to minimize loss of life and flood damage in areas downstream of the reservoirs. The degree of success in meeting these objectives depends primarily on two parameters: the lead time of the forecast, i.e. time lapse between the issue of a forecast and occurrence of flood peak, and the accuracy of the forecast, expressed in terms of the deviation between forecast and observed flood hydrograph.

These two parameters, lead time and accuracy, are interrelated in that a greater accuracy can be achieved for short lead times while long lead times are usually associated with inaccurate forecasts. What is desired, however, is a long lead time along with high accuracy, which can usually be obtained only to a certain degree by making compromises.

In small or mesoscale river catchments (like most rivers in Europe) the lead time of a forecast based on observed runoff data alone becomes very short, often so short that a forecast does not lead to any significant benefit. Several different approaches have therefore been tried to extend the lead time, particularly the techniques using rainfall-runoff models for flood forecasting, thus gaining the time of the transformation process. The use of radar rainfall measurements has the additional advantage of obtaining areal instead of point measurements, and of having total rainfall information available at a single point.

Often the use of rainfall-runoff models (R–R models) does not increase the lead time sufficiently. In such cases a QPF (quantitative precipitation forecasting) approach may be used which, however, leads to flood forecasts of low accuracy.

In this chapter we present a real-time flood forecasting technique which uses:

1. radar rainfall measurements as input into a R–R model;
2. a distributed system type R–R model in order to make use of the high resolution in space of the rainfall data (measured by radar);
3. an adaptive parameter optimization technique which optimizes the R–R model parameters each time a new forecast is issued (e.g. every hour) on the basis of the new data (rainfall and runoff) observed during the last time increment.
4. a stochastic real-time QPF technique to forecast the future rainfall until the end of the event while it is still raining.

These forecasts are accompanied by a specified probability of non-exceedance.

RAINFALL-RUNOFF MODEL

The rainfall is transformed into runoff with the aid of deterministic R–R model of the linear distributed system type. The catchment area is subdivided in area elements according to a polar co-ordinate system, the centre of which is formed by the radar. The chosen grid cell size is approximately 1 km^2 (1 km × 1 degree azimuth), as can be seen from Figure 13.1. For each grid cell the runoff computation is carried out in two consecutive steps:

    pure translation;
    storage attenuation effect.

The first step computes translation on the basis of time of concentration formulae for each grid cell, thus yielding a time–rain–concentration curve. In the second step the storage effects are computed with the aid of a transfer function which transforms the time–rain–concentration curve into a runoff hydrograph. Details of the so-called HYREUN model are published elsewhere (Anderl *et al.*, 1976; Schultz, 1969; Klatt and Schultz, 1985).

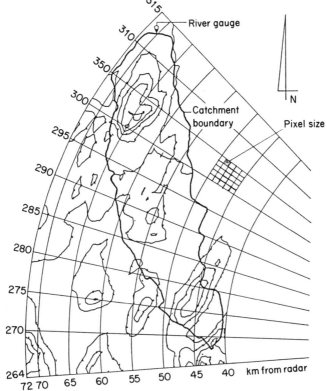

Figure 13.1. Subdivision of catchment into grid cells and isohyets (based on radar measurements)

RADAR RAINFALL DATA AS MODEL INPUT

A C-band radar of the German Weather Service located on an isolated mountain in Bavaria was used for rainfall measurement. As a test catchment the river Günz basin (a tributary to the Danube River) was used at Lauben (catchment area 318 km²) and Nattenhausen (526 km²) where the river flow is measured. The raw areal precipitation data are computed with the aid of the well-known radar equation (Attmannspacher, 1976). The raw rainfall depth values above representative surface raingauges are compared with the observed data from those gauges, and a calibration factor is calculated. Finally, all raw data are multiplied by the calibration factor.

The spatial resolution of the digital radar data was chosen to be the same as the grid cells of the R–R model. Time resolution was chosen appropriate to the catchment size and the flow times, giving intervals of $\Delta t = \frac{1}{4}$ h, $\frac{1}{2}$ h and 1 h. In Figure 13.1 $\frac{1}{4}$-hourly isohyets can be seen. The radar sampling time (horizontal beam) is every 5 min, thus $\frac{1}{4}$-hourly values contain the information from three measurements. This seemed necessary since it was shown that the accuracy of radar rainfall data decreases considerably if time intervals $\Delta t > 7$ min were chosen. If bright-band effects are not considered, the accuracy of radar measurements is approximately ± 15 per cent. For an area of 100 km radius a sensitivity analysis showed that three raingauges were needed for real-time calibration. This comparatively high accuracy (± 15 per cent) is also due to the fact that a different ZR relationship is used for different precipitation types.

Until now, only storms which did not contain hail or snow have been analysed; neither were events with bright-band effects taken into account. At present techniques are being developed which will allow the use of radar data under such conditions. Ground clutter was not a severe obstacle to the radar measurements in the test catchment areas.

REAL-TIME ADAPTIVE MODEL PARAMETER OPTIMIZATION

The HYREUN model contains several parameters, of which the two infiltration parameters are the most relevant for real-time flood forecasting. Comparatively small errors in these parameters result in significant over- or underestimation of the forecast flood hydrographs. Therefore the forecasting procedure discussed here contains an adaptive optimization procedure for these two parameters which can be applied in real time, i.e. during the flood forecasting computation.

The excess rain hyetograph is computed with the aid of a time-varying runoff coefficient: ψ

$$r_{\text{eff}} = r_{\text{obs}} \cdot \psi_{\text{end}} \left(1 - e^{-at} \right) \qquad (13.1)$$

Where:

r_{eff}	= excess rain;
r_{obs}	= observed rainfall, i.e. rainfall on a grid cell based on radar measurement during Δt;
ψ_{end}	= final runoff coefficient;
a	= constant;
t	= time.

Baseflow is chosen as a linear, slightly increasing function.

In real-time flood forecasting the first flood hydrograph forecast computation is based on the average values of the two relevant infiltration parameters Ψ_{end} and a, obtained from historical flood events. If a deviation between a computed forecast flood hydrograph and the observed hydrograph (observed until the time of forecast) is evident, it is assumed that this difference is due to false actual parameter values. This means that the initial parameters (Ψ_{end} and a from historical floods) have to be improved such that the deviation between the observed and computed (forecast) hydrograph becomes minimal. This optimization problem is solved using an objective function which minimizes the sum of squares of deviations:

$$\text{Min } C_v = \frac{\sqrt{\dfrac{\sum\limits_{i=1}^{n} (Q_{\text{obs},i} - Q_{\text{comp},i})^2}{n-1}}}{\bar{Q}_{\text{obs}}} \tag{13.2}$$

In equation (13.2) the least-squares principle is expressed in terms of the coefficient of variation.

Two versions of computer programs for this adaptive parameter control are used:

automatic computer optimization;
interactive parameter improvement.

The latter approach is demonstrated in Figures 13.2 and 13.3. In Figure 13.2 the parameter values (from historical floods) were:

$$a = 0.212; \quad \Psi_{\text{end}} = 0.11$$

After three successive improvements carried out by the person issuing the forecast, the solution of Figure 13.3 ($a = 0.27$; $\Psi_{\text{end}} = 0.14$) was found which shows a much better agreement between the rising limbs of forecast and the observed hydrograph. Later, when the total flood hydrograph had been observed, it could be shown that this solution was satisfactory not only for the rising hydrograph limb but also for the total hydrograph.

FLOOD FORECASTS USING QPF AS INPUT

As mentioned in the introduction, for small and mesoscale catchments the lead times of flood forecasts are usually too short for practical purposes such as flood warning and operation of flood retention storages. This is true even if the forecast is based on observed rainfall, e.g. by radar. Therefore the only way to extend the lead time further is the computation of quantitative precipitation forecasts (QPF). There are various techniques for QPF available—deterministic as well as stochastic methods—all of which contain a high degree of uncertainty. One of the most promising deterministic techniques is

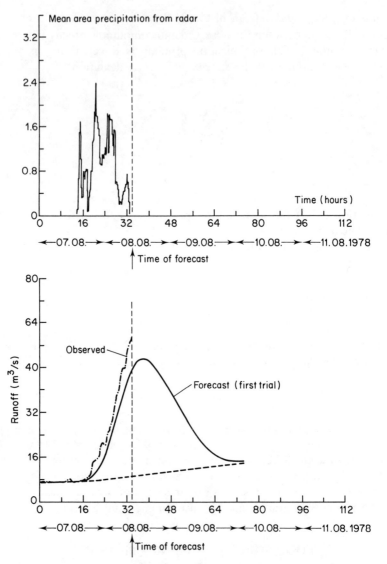

Figure 13.2.   Real-time flood forecast based on radar rainfall measurement

the FRONTIERS project, which combines radar with satellite data in order to compute a precipitation forecast.

In this chapter we have chosen a stochastic approach for forecasting rainfall. The depth and duration of the future rainfall is computed while it is still raining until the end of the storm event. The computed values are

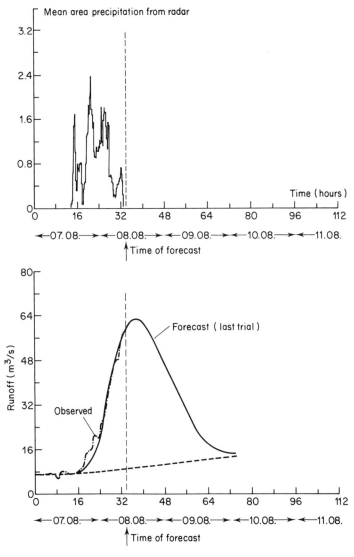

Figure 13.3. Real-time flood forecast based on radar rainfall measurement

given along with a prespecified conditional probability of non-exceedance. The condition is the quantity of rainfall which occurred from the beginning of the storm event until the time of forecast.

This estimation of future rainfall on the basis of observed rainfall is, in the sense of mathematical statistics, a conditional probability problem. It has

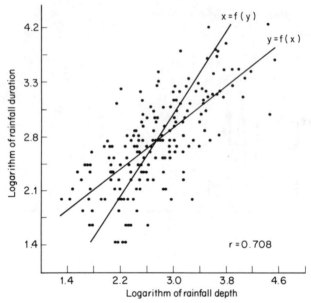

Figure 13.4. Linear regression between logarithms of rainfall depth and rainfall duration

n dimensions, n being the number of parameters describing future rainfall.
If the forecast problem is reduced to two parameters, i.e. future duration
and future amount of rainfall, then a conditional probability problem of two
dimensions has to be solved.

In the research project described here, 192 flood-producing rainfall events
of a special rainfall type (according to Aniol, 1971) were analysed. For this
rather large sample size a linear relation between the logarithms of duration
and depth of precipitation was established. The results of the regression
analysis are shown in Figure 13.4.

The correlation coefficient is $r = 0.708$. Although this correlation is not
too satisfactory, it can be used to reduce the forecasting problem to a one-
dimensional conditional probability problem. In this way the rainfall duration
can be expressed in terms of rainfall depth and vice-versa. Therefore it is
sufficient to estimate, for example, future rainfall depth and determine the
corresponding duration with the aid of the regression curve given in Figure
13.4.

The actual computation of future rainfall depth uses the conditional probability distributions (of rainfall depth) given the rainfall quantity already
measured. The most appropriate distributions were the Pearson-III distribution and the gamma distribution, which is a special case of the Pearson-
III distribution. Based on these probability distributions, future rainfall depth

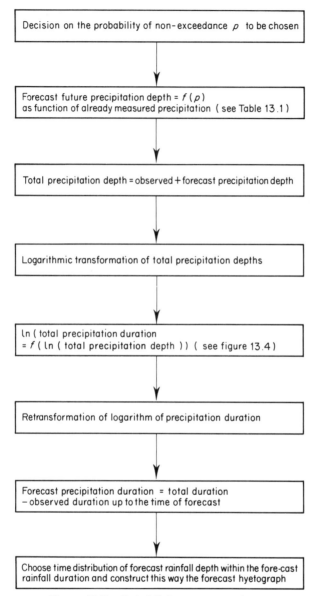

Figure 13.5. Rainfall forecast procedure

can be computed for any specified probability of non-exceedance, p (%). Table 13.1 gives an example of this dependence which is valid in Bavaria for a special type of rainfall. The complete rainfall forecast procedure is shown in the flow chart of Figure 13.5.

Table 13.1. Expected future rainfall depth for given observed rainfall and probability p of non-exceedance (example for southern Germany and special type of rainfall)

Observed rainfall depth (mm) $p(\%)$	Expected future rainfall depth (mm)										
	50	55	60	65	70	75	80	85	90	95	99
2	12.2	13.6	15.1	16.9	19.0	21.5	24.7	28.8	34.8	45.5	71.8
4	10.3	11.7	13.2	15.0	17.1	19.6	22.8	26.9	32.9	43.6	69.9
6	9.5	10.8	12.3	14.1	16.2	18.7	21.8	26.0	32.0	42.7	69.3
8	8.2	9.6	11.1	12.9	15.0	17.5	20.7	24.8	30.9	41.7	68.3
10	8.0	9.4	11.0	12.8	14.9	17.5	20.7	25.0	31.1	41.9	68.7
12	8.4	9.8	11.5	13.3	15.6	18.2	21.5	25.8	32.0	43.0	69.8
14	8.2	9.7	11.3	13.2	15.4	18.0	21.3	25.6	31.9	42.8	69.5
16	8.6	10.1	11.8	13.7	16.0	18.7	22.0	26.4	32.7	43.7	70.7
18	9.0	10.6	12.3	14.3	16.6	19.3	22.7	27.2	33.5	44.7	71.8
20	9.5	11.0	12.8	14.8	17.2	20.0	23.4	27.9	34.4	45.6	72.9
22	9.9	11.5	13.3	15.4	17.8	20.6	24.1	28.7	35.2	46.6	74.0
24	10.3	12.0	13.8	15.9	18.3	21.2	24.8	29.4	36.0	47.5	75.1
26	10.7	12.4	14.3	16.4	18.9	21.8	25.5	30.1	36.8	48.4	76.2
28	11.2	12.9	14.8	17.0	19.5	22.5	26.1	30.9	37.6	49.3	77.3
30	11.6	13.4	15.3	17.5	20.1	23.1	26.8	31.6	38.4	50.2	78.3
32	12.0	13.8	15.8	18.1	20.7	23.7	27.5	32.3	39.2	51.1	79.4
34	12.5	14.3	16.3	18.6	21.2	24.3	28.1	33.0	40.0	51.9	80.4
36	12.9	14.8	16.8	19.1	21.8	24.9	28.8	33.7	40.7	52.8	81.5
38	13.3	15.2	17.3	19.7	22.4	25.6	29.4	34.5	41.5	53.7	82.5
40	13.8	15.7	17.8	20.2	22.9	26.2	30.1	35.2	42.3	54.5	83.5
42	14.2	16.2	18.3	20.7	23.5	26.8	30.8	35.9	43.0	55.4	84.5

44	14.6	16.6	18.8	21.3	24.1	27.4	31.4	36.6	43.8	56.2	85.5
46	15.1	17.1	19.3	21.8	24.7	28.0	32.0	37.3	44.6	57.1	86.5
48	15.5	17.6	19.8	22.3	25.2	28.6	32.7	37.9	45.3	57.9	87.4
50	16.0	18.0	20.3	22.9	25.8	29.2	33.3	38.6	46.1	58.7	88.4
52	16.4	18.5	20.8	23.4	26.3	29.8	34.0	39.3	46.8	59.6	89.4
54	16.8	19.0	21.3	23.9	26.9	30.4	34.6	40.0	47.5	60.4	90.3
56	17.3	19.4	21.8	24.5	27.5	31.0	35.2	40.7	48.3	61.2	91.3
58	17.7	19.9	22.3	25.0	28.0	31.6	35.9	41.4	49.0	62.0	92.2
60	18.2	20.4	22.8	25.5	28.6	32.2	36.5	42.0	49.7	62.8	93.2
62	18.6	20.8	23.3	26.0	29.1	32.8	37.1	42.7	50.5	63.6	94.1
64	19.0	21.3	23.8	26.6	29.7	33.4	37.8	43.4	51.2	64.4	95.0
66	19.5	21.8	24.3	27.1	30.3	33.9	38.4	44.0	51.9	65.2	96.0
68	19.9	22.2	24.8	27.6	30.8	34.5	39.0	44.7	52.6	66.0	96.9
70	20.4	22.7	25.3	28.1	31.4	35.1	39.6	45.4	53.3	66.8	97.8
72	20.8	23.2	25.8	28.7	31.9	35.7	40.3	46.0	54.1	67.6	98.7
74	21.3	23.7	26.3	29.2	32.5	36.3	40.9	46.7	54.8	68.4	99.6
76	21.7	24.1	26.8	29.7	33.0	36.9	41.5	47.4	55.5	69.1	100.5
78	22.1	24.6	27.3	30.2	33.6	37.5	42.1	48.0	56.2	69.9	101.4
80	22.6	25.1	27.8	30.8	34.1	38.0	42.7	48.7	56.9	70.7	102.3
82	23.0	25.5	28.3	31.3	34.7	38.6	43.3	49.3	57.6	71.5	103.2
84	23.5	26.0	28.8	31.8	35.2	39.2	43.9	50.0	58.3	72.2	104.1
86	23.9	26.5	29.2	32.3	35.8	39.8	44.6	50.6	59.0	73.0	105.0
88	24.4	26.9	29.7	32.8	36.3	40.3	45.2	51.3	59.7	73.7	105.8
90	24.8	27.4	30.2	33.3	36.9	40.9	45.8	51.9	60.4	74.5	106.7
92	25.2	27.9	30.7	33.9	37.4	41.5	46.4	52.5	61.0	75.3	107.6
94	25.7	28.3	31.2	34.4	37.9	42.1	47.0	53.2	61.7	76.0	108.4
96	26.1	28.8	31.7	34.9	38.5	42.6	47.6	53.8	62.4	76.8	109.3
98	26.6	29.3	32.2	35.4	39.0	43.2	48.2	54.5	63.1	77.5	110.2
100	27.0	29.7	32.7	35.9	39.6	43.8	48.8	55.1	63.8	78.3	111.0

The analysis of all 192 observed rainfall events revealed that for a probability of non-exceedance, $p = 0.6$, the mean deviation between forecast and eventually observed rainfall of all the events was equal to zero. This therefore suggests using for real-time forecasts a value of p of at least 0.6. The decision as to which probability of non-exceedance to choose depends on the problem at hand, and on the risk-aversion or risk-proneness of the decision-maker.

For flood warning purposes an underestimation of the flood may be most dangerous, while overestimation—within reasonable limits—would cause no additional damage. Thus a high probability of non-exceedance should be chosen. On the other hand for the operation of flood protection storages overestimation may be very disadvantageous, since an unnecessary artificial flood may be created by reservoir operation on the basis of a forecast considerably overestimating a flood. In this way the operator may become liable for the damages caused by his reservoir operating procedure, and in such a case a probability of non-exceedance around 0.6 would be reasonable.

RESULTS

The success of the technique described above was verified using data from the Günz river catchment in southern Germany. The weather radar used was the C-band radar of the German Weather Service at the Hohenpeissenberg in Bavaria (Klatt, 1983). Figure 13.6 shows an example of a flood forecast after the end of the rainfall event compared to the observed hydrograph. The forecast shows a satisfactory accuracy. The peak and time-to-peak are in good agreement, while the flood volume is underestimated by about 10 per cent. This example shows the performance of an adequate rainfall-runoff model used for flood forecasting. The result is not surprising, however, since the complete rainfall information is available at the time of forecast and almost the whole rising limb of the actual flood is known, and thus used to adjust the model parameters.

Considerably more difficult and less accurate is a forecast issued before the end of the causative rainfall, since in this case less information about the actual rainfall and the actual flood hydrograph is available. Therefore, a rainfall forecast is required and the resulting flood forecast is dependent on the chosen probability of non-exceedance of the future rainfall. With the aid of the probabilistic model described above, future rainfall is estimated according to its depth and duration and the chosen probability of non-exceedance. However, in order to serve as input into a rainfall-runoff model, the time distribution of rainfall intensity within the duration must also be specified. Since no information about the future rainfall distribution is available, a realistic assumption is required. One possible assumption is to expect rainfall of constant intensity.

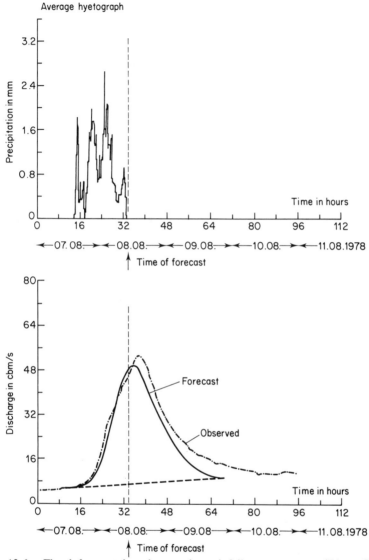

Figure 13.6. Flood forecast based on radar rainfall measurement (River Gauge Lauben, Günz River) after the end of rainfall

Figure 13.7 shows an example of a flood forecast while the rain is still falling, using this assumption about future rainfall. In the diagram four different examples of possible forecasts are given: (a) no future rainfall, (b) future rainfall with $p = 0.60$, (c) future rainfall with $p = 0.90$, (d) future rainfall with $p = 0.99$. It is obvious that even for $p = 0.99$ the flood peak

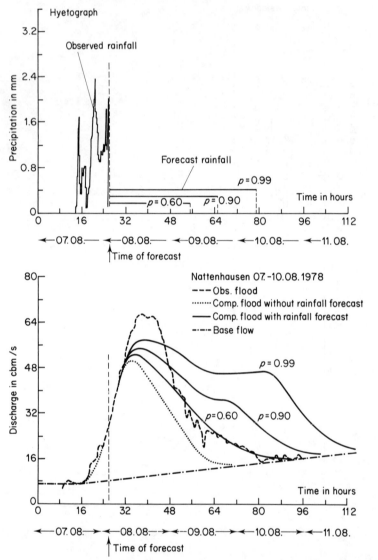

Figure 13.7.   Flood forecast based on radar rainfall measurement and rainfall forecast

is underestimated. The duration of the total flood is overestimated for $p >$ 0.6. This negative result is due to the unrealistic assumption about future rainfall intensity distribution. Figure 13.8 shows the same conditions except that the rainfall intensity is assumed to be linearly decreasing. The resulting forecast flood hydrograph is much more accurate than the one shown in Figure 13.7.

Unfortunately, up to now there are only a few radar-observed floods available for testing this, and it is not possible to present a statistical analysis of the accuracy of the technique.

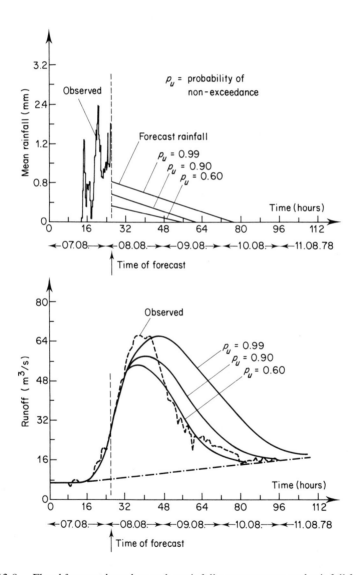

Figure 13.8.   Flood forecast based on radar rainfall measurement and rainfall forecast

RESERVOIR OPERATION ON THE BASIS OF FORECAST FLOOD HYDROGRAPHS

In addition to flood warning purposes, forecast flood hydrographs become increasingly important for real-time operation of flood protection storages. These exist either in the form of single-purpose flood retention reservoirs or as flood storage volumes in the upper zone of multi-purpose reservoirs. The whole procedure then consists of five consecutive steps:

1. Radar rainfall measurements in real time and calibration of raw data.
2. Computation of a precipitation forecast, i.e. future rainfall depth and duration until the end of the event while it is still raining (for various probabilities of non-exceedance).
3. Application of a rainfall-runoff model using observed and forecast rainfall as input and computing the expected (with various probabilities of non-exceedance) flood hydrographs as output (= forecast flood hydrograph).
4. Computation of an optimum reservoir operating policy during the flood in real time on the basis of the forecast flood hydrographs.
5. Activation of reservoir gates in order to implement the optimized reservoir operating policy.

Items 4 and 5 have not been dealt with here, but in passing it should be noted that various objective functions have been developed for the optimization procedure (Schultz and Plate, 1976). The major problem here, however, is the effect of the flood forecast uncertainty on the operation of a reservoir system; and it is in this field (objective functions, effect of forecast uncertainty) that further research is needed.

The Federal Republic of Germany is participating in the COST Project (see Chapter 2) concerning the establishment of a West European weather radar network. It is hoped that in about a decade from now all flood protection storages could be operated on the basis of flood forecasts using radar rainfall measurements. The expected use of satellite data for improved forecasts and a more efficient application of operations research techniques (including uncertainty of forecasts) should enable us to operate flood storages in the future more efficiently to reduce the loss of life, as well as damage due to floods.

ACKNOWLEDGEMENTS

The author wishes to thank Dr W. Attmannspacher of the German Weather Service for providing the radar rainfall data, and the Bavarian Institute of Water Resources for providing the runoff data. Thanks are also due to the German Research Foundation for financial support of the research work.

The computations were carried out at the Computer Centre of the Faculty of Civil Engineering of the Ruhr-University Bochum.

REFERENCES

Anderl, B., Attmannspacher, W., and Schultz, G. A. (1976). Accuracy of reservoir inflow forecasts based on radar rainfall measurements. *Wat. Resour. Res.* **12**(2), 217–23.

Aniol, R. (1971). Sommerniederschlag am Hohenpeissenberg und Wetterlage, Sonderbeobachtungen des meterologischen *Observatoriums Hohenpeissenberg*, no. 17.

Attmannspacher, W. (1976). Radarmessungen zur Bestimmung von Flächenniederschlägen. *Die Naturwissenschafen,* **63**, 313–18.

Klatt, P. (1983). Vorhersage von Hochwasser aus radargemessenem und prognostiziertem Niederschlag. *Schriftenreihe Hydrologie/Wasserwirtschaft,* No. 3 (ed. G. A. Schultz), Bochum.

Klatt, P., and Schultz, G. A. (1985). *Flood forecasting on the basis of radar rainfall measurements and rainfall forecasting.* IAHS publ. no. 145, pp. 307–15.

Schultz, G. A., and Plate, E. (1976). Developing optimal operating rules for flood protection reservoirs. *J. Hydrol.,* **28**(3), 245–64.

Schultz, G. A. (1969). Digital computer solutions for flood hydrograph prediction from rainfall data. In: *The Use of Analog and Digital Computers in Hydrology. Proc. Tucson Symp.,* 1968. IAHS publ. no. 80, pp. 125–37.

Part IV

WEATHER RADAR TECHNOLOGY IN THE FUTURE

In this concluding section the authors look to the future in terms of radar networks and data use. The ways by which hydrological modelling for flood forecasting might develop as radar data becomes more widely available are discussed in Chapter 15, with wider possibilities for the use of radar data by the water industry being considered in Chapter 14. The last two chapters, written by senior Meteorological Office staff, look to future developments of the UK radar network and to use of data from this network and from satellites for contributing to improved weather forecasting.

Weather Radar and Flood Forecasting
Edited by V.K. Collinge and C. Kirby
© 1987 John Wiley & Sons Ltd.

CHAPTER 14

The Role of Radar and Automated Data Capture in Information Systems for Water Management

P. D. WALSH AND A. M. LEWIS

INTRODUCTION

The North-West Weather Radar Project (NWRP) has shown that real-time telemetry of river flows and rainfall from traditional sensors and from an unmanned weather radar can be integrated successfully into a viable system for meeting the hydrological information needs of river management. The requirements of a water authority for hydrological information, however, embrace the whole water cycle. Recent developments in information technology provide opportunities to make hydrological data more widely available to meet information needs which could benefit all aspects of water management.

The North West Water Authority is currently investing heavily in its strategies for communications and telemetry, into which the current NWRP will be incorporated. This will allow the development of an integrated system from 'sensor' to 'archive', embracing the needs for hydrological data to service all functions. Early benefit should occur from the wide range of applications to which short-term rainfall forecasts produced by FRONTIERS can be applied. In the longer term, development of database technology with associated fast search and retrieval facilities will allow data to be used more effectively in the planning, operation and monitoring aspects of water management.

In this chapter we postulate a concept for the future in which hydrological data collection and manipulation in real time is integrated with meteorological information as part of an efficient communication network, linking users directly to the data and associated archives. In turn, decision-making requirements, hitherto only partially satisfied, are more fully met by the integration of historical (archive) and real-time data systems.

The British water industry can boast a long history of collecting data about its primary activities, but the data collection programmes have often developed without proper regard to the needs for data and information. Furthermore, the industry has failed to capitalize upon its investment in collecting the data, with invaluable information being discarded or ignored because the data could not easily be adapted into its most useful form. In recent years the collection and processing of data for long-term time series 'archives' has been eclipsed by moves towards real-time monitoring and control systems with the more evident operational benefits. Nowhere is this more clearly demonstrated than for the flood warning function *vis-à-vis* river flow measurement.

Increasing emphasis on monitoring performance and on the provision of information for management focuses attention on the archival storage of a very wide spectrum of physical and financial data. Consequently, concepts of co-ordinated databases, spatial and graphical presentations, and user-friendly reports abound. Incentives to improve performance are therefore now leading to critical examination of the ways and means available for effective data handling and use.

It is into these developments of integrated water management information systems that the North-West Weather Radar Project, with its real-time collection of rainfall and river flows, must now be incorporated. This system has the ability to provide actual and forecast information for dissemination via telemetry and distributed computer networks to different users locally and regionally, as well as to produce data for the long-term archives. This information is required to service the wide range of activities performed by the Water Authority.

North West Water is responsible for all aspects of management of the water cycle over an area of 14 500 km². It supplies 2500 megalitres of drinking water every day after treatment at 200 water treatment plants. It needs to dispose of three million wet tonnes of sludge produced by 700 sewage treatment plants every year. In addition to the primary functions of water supply, sewerage and sewage treatment. North West Water, like all other regional water authorities in the UK, is responsible for all aspects of river basin management, including water quality, land drainage, fisheries and the flood warning service. It is a multi-million-pound industry which needs to invest heavily (currently £180 million per year) to improve or maintain its services at an acceptable level and to counteract the pollution inherited from the region's industrial heyday.

Table 14.1. Methods of dissemination

Speech
Text
Telex
Electronic mail
Visual display units
   Tabular
   Report
   Graphical
   Diagram
   Map

Private teletext service

Public teletext services
   Telephone-based (interactive)
   Television-based (dumb)

APPLICATIONS

Although the principal benefit of radar data within North West Water had to date been to flood forecasting, there are many other potential benefits over a wide range of applications such as river regulation and improved engineering construction and design. (National Water Council–Meteorological Office, 1983).

However, for the effective integration of the data collection system, and for the benefits of major developments in rainfall forecasting such as FRONTIERS (Browning, 1979 and Chapter 16 of this volume) to be realized, it is essential to establish how the data would be used in different applications (Krietzberg, 1981) since there is no inherent value in the data themselves; only the information that data convey to the decision process in which they are being used has value. Consequently, data must be presented in forms suited to the function in which it is being applied.

Data on the hydrological cycle can benefit most aspects of water management and society at large. Hydrological information has a myriad of uses from the local to the national level (Rodda and Flanders, 1985). The water management functions can be identified as: (a) water resources, supply, treatment and distribution; (b) sewage treatment and sewerage; (c) river quality; (d) land drainage; (e) recreation; and in some areas (f) navigation. Clearly there should be correspondence between these various functions and the information flows of the management system to meet the management objective of providing these services.

Therefore the task becomes one of dissemination by an appropriate method (see Table 14.1) and in a suitable format to enhance the management decision process. Inter-functional conflict should be avoided if recent developments in information technology are to be utilized in a comprehensive information strategy that covers data collection, telemetry, data processing, analysis and storage. The strategy should seek to link needs for technical information to those of finance and personnel.

How will hydrological information and knowledge be applied in the industry of the future? They will continue to be used to support decision-making and should, if fully utilized, improve the following three types of decision:

Control

These decisions are concerned with maintaining some state within specified margins. In the water supply function the status of a reservoir needs to be optimized to meet a range of objectives embracing cheap gravity supply to hydropower generation. At a sewage works the throughput and treatment processes must be controlled to maintain effluent standards with minimum energy consumption. In both these cases real-time application of hydrological data should lead to improvements and economies in operation. It has been shown, for example, that with sufficient knowledge of individual systems it is possible to use mathematical models for optimization of sewage pumping, leading to corresponding savings in electricity costs (Evans, 1981). These models were derived by the application of system engineering techniques that identified the inherent decision-making process and the type of data needed to give the requisite information, i.e. real-time rainfall and water level data.

Resource allocation

These decisions are directed to carrying out some task in a given period of time. The proper allocation of resources leads to the attainment of an objective, such as achieving appropriate levels of reservoir refill during winter months. In the context of water supply systems allocation of demand to different sources, e.g. groundwater, river, reservoir, can be optimized in operational planning studies, leading to control and operational rules for the system. The former require 'archival' hydrological information whilst the latter would use real-time information on rainfall and river flow. In addition the actual application of the data requires accurate knowledge of the processes in order to achieve sound hydrological forecasting. Studies into applying data from the NWRP to operation of the Lancashire Conjunctive Use Scheme have shown that considerable work still needs to be done in this area in order to exploit the information content in the data more fully (Walker *et al.*, 1983).

Planning

These decisions are concerned with future actions, which involve defining new objectives or the better definition of existing ones, followed by the provision of adequate and appropriate resources to meet them. By their nature planning investigations require long-term data, which are often derived as a subset of short-term operational data, as a basic requirement for understanding the system. However as computer techniques for modelling systems are increasingly utilized, certain types of application demand calibration of a system where little or no long-term data exist. For example, the use of the Wallingford procedure for hydraulic analysis of sewerage systems in sewerage rehabilitation studies (Water Research Centre, 1983) frequently relies upon a special flow survey covering only a few weeks, and in order to calibrate sewerage systems, estimates of areal rainfall over urban areas are needed. One potential source of such rainfall data is that produced by the weather radar. Once calibrated the WASSP model (National Water Council, 1981) may then be used for the planning and design of sewerage rehabilitation works.

Many planning and scientific studies in the north-west are directed at the major task of improving surface water quality (NWWA, 1983). Rainfall data and forecasts linked to routing models for sewers have the potential to be used to improve the operational performance of stormwater overflows which could lead to reduction of pollution from the 'first flush' on the receiving water (Damant *et al.*, 1983).

INFORMATION NEEDS

Hydrological information can contribute to many areas of decision-making. More significantly, projects such as NWRP and NWWA telemetry provide the basis for a viable hydrological information system which can meet wider objectives than those for which they were originally conceived and currently serve. However, successful integration of this information system into a wider management information system demands careful examination of the water authority's information needs in the widest sense.

In a series of local studies covering different functions North West Water is attempting to define and elucidate information needs and the various data flow patterns for a particular function. To do this it is necessary to examine objectively the organization, its functions, purpose, resources (including manpower) and the environment in which it operates. Since water services are predominantly a process industry enveloped in and dependent upon a natural environmental system, this adds a dimension to its information needs not normally shared with other process industries. It cannot design or own the natural system and there are limited choices for its control. Thus the

Table 14.2. Conversion of site data to information

---

1. *Data collection from site*
   Satellite—data collection platform
   Radio
   Private land-line
   Telephone
   On-site logger

2. *Processing*
   Quality control
   Conversion, e.g. river level to flow rates to totals
   Archival storage

3. *Application/conversion to information*
   Forecasts (real time)
   Analysis (planning)

4. *Dissemination of useful information*
   For example: Flood warning
                  Pollution time of travel
                  Rainfall
                  Results of analysis

---

information required becomes that for understanding its behaviour, leading to an ability to predict future behaviour and therefore control the use of its resources.

It can be argued that for its principal purpose of providing a comprehensive water service a water authority must have data derived from measurements throughout its process and distribution facilities and from the natural world, none of which stop working in real time. The information from the systems is used both for on-line control and for gaining knowledge. Table 14.2 lists the major stages in the acquisition and conversion of data to usable information, and illustrates typical features of the four stages. For the natural system this requires the accumulation of long-term time-series data—the archival data. There is also a correspondence between the operational functions, e.g. water resources today are tomorrow's water supply and effluents and may then even become a future downstream source.

The requirements for information demand information systems that cover real-time operation and long-term archiving of material, with the latter, in particular, showing high integrity. In other words, systems of automated data capture must be capable of providing data at different frequencies and possibly different accuracies to meet a range of purposes and information needs.

It is unlikely that the future will see different networks for real-time data usage compared with long-term data as in the case of current river flow networks.

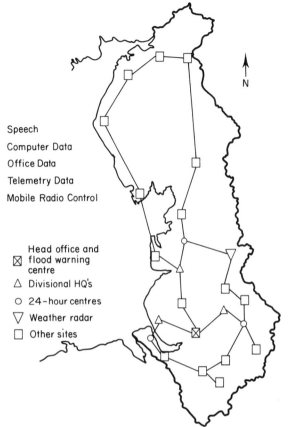

Speech
Computer Data
Office Data
Telemetry Data
Mobile Radio Control

⊠ Head office and
   flood warning
   centre
△ Divisional HQ's
○ 24-hour centres
▽ Weather radar
☐ Other sites

Figure 14.1.   North West Water integrated private
communications network

INFORMATION SYSTEMS

Given these arguments we need to look to a comprehensive system that comprises:

1. accurate and reliable sensors.
2. communication to appropriate reporting centres
3. front end processing and data manipulation for real-time operations,
4. quality control for archival material,
5. communication to archival locations,
6. retrieval system for data users.

It is into the component parts (1) and (2) that the Weather Radar Project can now be integrated as North West Water embarks on a comprehensive programme of telemetry and communications (Figure 14.1). The overall objective of the private communications network is to provide capacity for carrying securely most speech, computer data, office data, telemetry data and for mobile radio control between a large number of authority installations and operational centres over an area of 14 500 km².

For each functional district, telemetry systems are now being installed. A system will comprise a master station scanning outstations, collecting and collating data to produce operational reports in real time. All outstations will have sufficient local storage and intelligence to carry on operating independently of a master station, if the latter fails.

The individual master stations will in turn be linked by a 'data-ring' to district and divisional centres. Since some information is more 'regional' than local in character, current thinking suggests that this will require regional or subregional processing centres. These would be the focal point for the collection of regional information and provide analysis and reports for dissemination back down to local level, as well as passing it to long-term archives. River level, rainfall and FRONTIERS products would fall into this category. For hydrometric information the scanning of individual sensors will still be performed by local intelligent outstations, but the information will be routed to the regional centre as a precursor to its direct use; not least in the regional flood warning centre. Thus telemetry systems are an integral part of a communications network taking data from 'field' to 'archive', meeting information needs on route (Figure 14.2).

The concept is ambitious but the potential to give wider application of the captured data, and in particular regional data such as rainfall, is enormous. However, as with all systems there are problems to be solved before objectives are met. These range from adequate sensors to appropriate staff skills for software generation, the maintenance of the hardware system and for data management and control. (Institution of Water Engineers and Scientists, 1982).

It is generally recognized that sensors have been the weakest link in automated data capture systems. Indeed for some applications satisfactory sensors do not exist. The establishment by the Water Research Centre of evaluation and demonstration facilities for the testing of sensors was an important initiative (ICA Steering Group, 1985). In the USA the geological survey has also recognized that sensor development is an important task (Walsh, 1984).

Adequate maintenance of these systems is vital to their success. The NWRP has demonstrated that high returns of data can be achieved. Similar efforts will be required for all other elements of data acquisition systems. This has been examined in a separate study (NWWA, 1984) which concluded that systems engineering techniques should be employed to derive and control methods of preventive maintenance.

Finally the appropriate skills will need to be acquired by managers, engin-

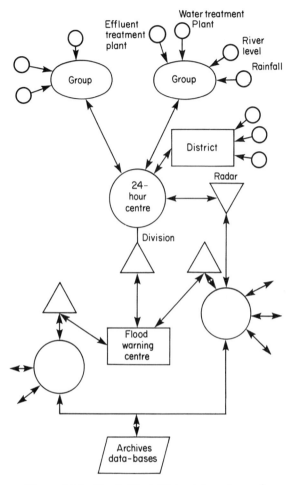

Figure 14.2 North-West Water telemetry and
communications

eers and scientists to manage and exploit the data and information. It will
not be appropriate for them to rely upon information and computer specialists
for detailed analysis and presentation. They must become at least as adept,
and possibly more so, with the presentation of numerical and graphical
information as with the spoken and written word.

The motivation to move North West Water forward in terms of information
technology, e.g. telemetry, automated data capture and regional communications, has come from a desire to provide a secure, cost-effective
service for its customers. It has not been promoted by a desire to collect
data for data's sake, nor by esoteric concepts of understanding catchment

and other processes in minute detail. Almost certainly there will be more rationalization of existing hydrometric data networks over the next few years. This will be balanced by an increase in measurement and analysis of other data types within the sewerage and distribution functions to aid better understanding of the processes and to enable their optimization by increased application of instrumentation, control and automation techniques. On the other hand care will have to be taken to ensure that long-term data requirements are adequately covered so that our understanding of natural processes, and our ability to detect changes in them, is not weakened.

CONCLUSIONS

For the future the picture is one of a traditional hydrometric data network utilizing new technologies to become more efficient and make data more widely available. Its use will be dictated by the information needs of functions within the water industry, which comprise operational processes operating in the natural environment. Therefore, there will always remain the necessity to collect accurate data in real time and collate such data into long-term information.

It is envisaged that single-purpose databases will cease to exist, but will be integrated onto a common database to which all functions will have access by means of fast retrieval techniques. As these techniques develop increasing emphasis will have to be placed on the best means of presenting data to the user for his individual usage.

Information is a scarce and expensive resource. Developments such as weather radar projects, communications and telemetry private networks could generate an information explosion. For effective use the information content and the data collection and application of these systems will need as much careful planning as the computer and telemetry hardware. As a resource, information requires and deserves effective management performed with the same dedication organizations apply to managing their other resources.

The data from NWRP and associated systems are available and ready to play their part in improving our management of the hydrological and water cycles, and should now be exploited to their full potential.

ACKNOWLEDGEMENTS

This chapter draws heavily on ideas discussed with many colleagues at North West Water, and the authors wish to acknowledge that assistance in developing their thinking. Although reference is made to a number of current developments at North West Water, this chapter is the author's view of a possible future and should not be interpreted as a statement of policies of the authority.

REFERENCES

Browning, K. A. (1979). The FRONTIERS plan: a strategy for using radar and satellite for very short-range precipitation forecasting. *Met. Mag.*, **108**, 161–84.

Damant, C., Austin, G. L., Bellen, A., Osseyrane, M., and Nguyen, N. (1983). Radar rain forecasting for wastewater control. *Proc. Am. Soc. Civ. Engrs*, **109** (HY 2), 293–7.

Evans, G. P. (1981). Optimisation of sewage pumping. Water Research Centre Technical Report TR170, Water Research Centre, Medmenham.

Institution of Water Engineers and Scientists (1982). Symposium on Management and Application of Micro-Electronics in the Water Industry. Institution of Water Engineers and Scientists, London.

Kreitzberg C. W. (1981). Flood information exchange aspects of the Integrated Flood Observation and Warning System (INFLOWS). Co-operative Agreement NA79AA-H-00108, prepared for R. L. Carnahan, Chief, Weather and Flood Warning Co-ordination Office, National Weather Service, Silver Spring, USA.

National Steering Group on New Technology and ICA (1985). *Instrumentation Control and Automation in the Water Industry*. Water Research Centre, Swindon, 15 pp.

National Water Council–Meteorological Office, (1983). Report of the Working Group on National Weather Radar Coverage, National Water Council, London, 31 pp.

National Water Council Department of the Environment (1981). Design and analysis of urban storm drainage—The Wallingford Procedure. NWC/DoE Standing Technical Committee Report No. 29. National Water Council, London.

North-West Water Authority (1983). Improving rivers, estuaries and coastal water in the north west. Consultation Paper, North-West Water Authority, Warrington, 13 pp.

North West Water Authority (1984). Report and guidelines on repair, maintenance and support of ICA systems. Draft report of NWWA ICA Working Party–Maintenance Sub-Group, North-West Water Authority, Warrington, 28 pp.

Rodda, J., and Flanders, A. (1985). The organization of hydrological services. In J. Rodda (ed.), *Facets of Hydrology II*. Chichester and New York: Wiley, 447 pp.

Walker, S., Nelson, J. A., and Austin, P. (1983). Assessment of savings in operational costs of Lune-Wyre transfers. North-West Weather Radar Research and Applications Project 4.2, North-West Water Authority, Warrington.

Walsh, P. D. (1984). Hydrometeorological data collection practices and management in the USA. Report on 1983 Churchill Fellowship Study Tour, North-West Water Authority, Warrington, 60 pp.

Water Research Centre (1983). *Sewerage Rehabilitation Manual*, Water Research Centre, Swindon.

Weather Radar and Flood Forecasting
Edited by V.K. Collinge and C. Kirby
© 1987 John Wiley & Sons Ltd.

CHAPTER 15

Towards More Effective Use of Radar Data for Flood Forecasting

R. J. MOORE

INTRODUCTION

In order to speculate on how weather radar may be put to more effective use for flood forecasting in the future, and on the degree of benefit likely to accrue, it is important to understand the major shortcomings of existing flood forecasting procedures. Comparison of flood forecasting models with different structural forms often leads to a result which is not clear-cut, with no one model consistently outperforming all others. A threshold of model performance appears to be reached by a range of rainfall-runoff models. The main key to surpassing this threshold of model performance is seen to be an improved estimation of areal rainfall; thus on *a priori* grounds the likely benefits of weather radar appear very promising indeed.

Scope for improvement using weather radar data appears to exist not only through better definition of the catchment average rainfall used as a lumped input to many rainfall-runoff models, but also in the better use of the spatially distributed rainfall measurements that weather radar provides. However, the use of spatially distributed measurements for flood forecasting first requires the development of rainfall-runoff models which are structured to make greater use of the information on spatial variation in rainfall contained in these data. The development of such models forms a central theme of this chapter. An outline of a simple geometrically distributed model is presented, together with a novel rainfall-runoff model, which treats rainfall as a pro-bability-distributed variable over the basin. Both models are designed to make better use of weather radar data. Updating techniques capable of employing flow information up to the present to compensate for the effect of past rainfall errors on rainfall-runoff model performance are considered: a scheme based on empirical adjustment of the water content of a conceptual model's stores is presented as an attractive alternative to more formal schemes based on the Kalman filter.

In the concluding sections the case is argued for a more rigorous evaluation than heretofore of the benefits of weather radar data to flood forecasting, carried out across a range of hydrological environments: such an evaluation is seen as a crucial requirement for the future.

INFLUENCE OF RAINFALL ERRORS ON RAINFALL-RUNOFF MODEL FORECASTS

Consideration needs to be given first to the effect on rainfall-runoff model performance of error-corrupted rainfall measurements, either in the form of at-a-point measurement errors or through the failure of a model to take account of the distributed nature of the rainfall. This serves to clarify both the potential value of radar-derived measurements of rainfall to flood forecasting, and the implications of their error-corrupted nature to rainfall-runoff model design. It is important to understand the influence of these errors on rainfall-runoff model performance if an appreciation of the potential value of distributed radar rainfall data is to be gained. The following examines the relation of model structure to rainfall input errors, and the error introduced by using a lumped, areal average, rainfall input.

Effect of the structural form of the model

The effect of errors on rainfall measurements used as input to a rainfall-runoff model will depend on the model's structural form. Singh and Woolhiser (1976) demonstrate how a non-linear rainfall-runoff model tends to amplify the errors so that a linear model calibrated on data simulated from a non-linear model and using exact rainfall may perform better than the true non-linear model when the rainfall input is error-corrupted. Singh (1977) demonstrates that even if rainfall is exactly known, the errors introduced in conversion to rainfall excess may mean that a linear routing model outperforms a non-linear one. The simulation results of Singh and Woolhiser provide a persuasive reason for using linear models; however, experience using actual data suggests that in many circumstances non-linear models provide improved performance, presumably because amplification of errors by non-linear models is less deleterious than the use of a linear model to represent the non-linear rainfall-runoff process in these situations. Weather radar's shortcoming in not always providing reliable quantitative measurements of rainfall (Collier *et al.*, 1983), and the need for robust flood forecasts, may mean that linear models might be preferable in some cases.

Effect of using lumped, areal average, rainfall

Basin average rainfall estimated by a single raingauge measurement or as an average of measurements from a small number of raingauges will have a greater variance than rainfall averaged over an infinite number of points within the basin. This error associated with raingauge network-derived estimates of basin average rainfall may be termed a space-sampling error and is a major cause of bias in runoff prediction (Troutman, 1982, 1983). Its effect is to tend to make the rainfall-runoff model overpredict large events and underpredict small events. In practice this bias is compensated for by calibrating the rainfall-runoff model by least-squares using the raingauge network-derived areal estimates of rainfall. However, if a model is calibrated in this way, and then used operationally for flood forecasting using radar-derived areal average rainfalls, there will be a tendency to underpredict large events and overpredict small events, on account of the lesser variability of the radar-derived estimate. This serves to emphasize the importance of calibrating rainfall-runoff models using the data source which is to be used in real time. In the event of a weather radar malfunctioning, it would be desirable to switch to the use of telemetering raingauges and a rainfall-runoff model specifically calibrated using data from these gauges. Ideally a flood warning system should accommodate a range of model calibrations appropriate to each possible configuration of raingauge and radar measurements of rainfall, thereby catering for all possible scenarios of data availability.

THE POTENTIAL VALUE OF WEATHER RADAR DATA

Moore (1977) investigated the effect of using several raingauges located in and around a basin as inputs to a multiple-input transfer function model; data from up to six raingauges in the vicinity of the small (33.9 km²) Hirnant basin in North Wales, were employed. The inclusion of more than two gauges, either as separate inputs or as a single average value, failed to improve forecasts of hourly flows. It was argued that a gauge, or set of gauges, should not be chosen for use in flood forecasting in terms of how well it estimates the basin average rainfall, but rather on the strength of its association with observed river flows. Siting of a single gauge in the vicinity of the contributing area of storm runoff (for example, located near the basin outlet) could be superior for flood peak forecasting than a dense network designed to obtain a good estimate of the basin average rainfall. This argument can be forwarded to suggest that only radar grid square estimates of rainfall located in the area contributing to the flood hydrograph rising limb and peak be used for flood forecasting. Observed flows may be used in a forecast updating algorithm to compensate for forecast errors on the falling limb, where flows originate from rain falling on more distant parts of the basin. Research leading to a prescription for rainfall measurement requirements for real-time flood forecasting systems, which takes into account the complementary role of radar-derived and gauge estimates of rainfall, is seen as a need for the future.

Storm movement

Weather radar data may be of most value when storm movement across a basin exerts a strong influence on the form of the resulting flood hydrograph. Hamlin (1983) presents the results of Bramley's work on simulating a storm hydrograph by moving a fictitious storm cell across a basin. The effect of sampling rainfall from the storm cell by a varying number of point rain-gauges is investigated and shown to result in large errors in the timing and volume of the hydrograph peak. In such situations the rainfall pattern obtained from weather radar could be very useful in ameliorating the errors incurred by inadequate sampling of the rainfall field by a sparse rain gauge network. However, this would depend on the availability of a distributed rainfall-runoff model to fully utilize the weather radar data, and on the accuracy of the weather radar data. Ngirane–Katashaya and Wheater (1985) also demonstrate the importance of storm movement on flood peak generation through simulation experiments on a synthetic urban catchment. Niemczynowicz (1984), in a comprehensive investigation of the areal and dynamic properties of rainfall and its influence on runoff production, illustrates the importance of catchment shape and orientation relative to the direction of storm movement: greater peak discharges result from long, narrow basins orientated in the prevailing storm direction. Weather radar data may therefore be expected to be of greatest value on such basins, if used as input to a model which is distributed in form; the development of such models will now be discussed.

DISTRIBUTED MODELS FOR REAL-TIME FLOOD FORECASTING USING GRID SQUARE WEATHER RADAR DATA

With the increasing availability of radar-derived rainfall data for use in operational flood forecasting, the hydrologist has an obligation to make the best use of such data through the development of distributed models capable of real-time implementation. Such models would have the potential to overcome the poor performance of lumped rainfall-runoff models in situations where spatial non-uniformity of rainfall and storm movement are important, as previously discussed. Whilst physics-based models (Abbott *et al.*, 1978; Morris, 1980) are distributed in three dimensions, with changes in water

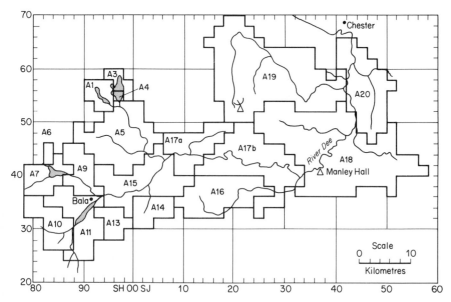

Figure 15.1. Grid square representation of the River Dee subcatchments used for radar purposes (each grid is 4 km² in area)

content with depth being modelled through solution of the equations of Richards and Boussinesq, there is scope for development of simple distributed models which lump the depth dimension, for example into soil- and groundwater components; these models should be capable of implementation in real-time on a continuous basis whilst making better use of the distributed radar-derived rainfall data. The most common approach is to subdivide a basin into subcatchments and channel segments, and use lumped rainfall-runoff models to forecast subcatchment flows. This approach was used as part of the Dee Weather Radar Project (Central Water Planning Unit, 1977). Figure 15.1 shows how the subcatchments were approximated by the 2 km radar grid in order to derive subcatchment average rainfalls from the radar grid square values. Considerable loss of information on spatial variability may be incurred depending on the relative magnitudes of subcatchments and radar grid squares, and the degree of rainfall variability experienced over the subcatchment. A simple, alternative scheme would be to represent each radar grid square by some form of transfer function or non-linear storage model, and obtain the basin flow response by combining their individual responses, after being shifted in time in relation to their distance from the basin outlet (possibly also including slope information to better approximate time of travel). A similar scheme could be used to forecast lateral inflow to a channel flow routing model. An example of where a grid square model has been used with radar data is presented by Anderl *et al.* (1976): an effective rainfall separation and time shift at each grid square is followed by routing of the combined flows through two unequal reservoirs. This model is shown to provide more accurate forecasts with radar data as input than the use of spatially interpolated estimates, derived using data from a regular raingauge network. The value of adopting a more distributed model would, of course, depend on the number of radar grid squares encompassed by the basin, and might be expected to be of greatest value in parts of the world where flood forecasts are required for basins larger than those found in the United Kingdom.

A probability-distributed model of rainfall and runoff production for use with weather radar data

Instead of developing a geometrically distributed model of basin runoff, an attractive compromise would be to utilize the information on spatial variability of rainfall provided by weather radar to define a spatial distribution function of rainfall, and to incorporate this into a rainfall-runoff model. Thus the mean rainfall used in a lumped model is replaced by a spatial distribution of rainfall, characterized not only by its mean but by the form of variability about this mean. Whilst the geometric pattern of rainfall over space is discarded in this compromise approach, information on the frequency of occurrence of rainfall of given magnitudes over the catchment is retained. We will now discuss how this spatial frequency information may be incorporated into a rainfall-runoff model.

Consider that runoff production at a point is controlled by a simple excess mechanism, so that rainfall above a given value becomes runoff. This excess mechanism may be due to the soil not allowing water to enter above some maximum rate (the infiltration capacity), or due to the soil having a limited capacity to store water (the storage capacity). Mathematically this may be represented by

$$q = \begin{cases} p - c & p > c \\ 0 & p \leq c \end{cases} \tag{15.1}$$

where c is the threshold capacity, p the rainfall, and q the runoff. The spatial variation of rainfall and threshold capacity over the basin may be considered as a bivariate probability density function, $f(p,c)$, such that $f(p,c)dpdc$ denotes the probability of rainfall and capacity at any point in the basin being in the range $(p,p+dc)$ and $(c,c+dc)$. It is reasonable to assume that p and c are independent random variables with density functions $f_p(p)$ and $f_c(c)$, so that $f(p,c) = fp(p) f_c(c)$; $\bar{p}$ and $\bar{c}$ will be used to denote the mean rainfall (the expected value of p) and the mean capacity respectively.

According to the point description of runoff production specified by equation (15.1) it follows that runoff production from the entire basin will be given by

$$Q = \int_0^\infty \int_c^\infty (p-c) f P(p) f_c(c) \, dp \, dc \ .$$ (15.2)

Some algebra leads to the result

$$Q = \bar{p} - \int_0^\infty (1 - FP(p))(1 - F_C(p)) \, dp \ ,$$ (15.3)

where $F_p(p)$ is the distribution function of rainfall, indicating the proportion of the basin where rainfall is less than p, and $F_C(p)$ is the distribution function of capacity, indicating the proportion of the basin where the capacity of the soil to take up water is less than p. For ease of illustration, we will assume the density functions to be exponential, so that:

$$F_P(p) = 1 - \exp(-p/\bar{p})$$ (15.4a)

$$F_C(p) = 1 - \exp(-p/\bar{c}) \ .$$ (15.4b)

Then:

$$Q = \bar{p} - \int_0^\infty \exp\left\{ -p\left(\frac{1}{\bar{p}} + \frac{1}{\bar{c}} \right) \right\} dp$$ (15.5)

which leads to the simple relation

$$Q = \frac{\bar{p}^2}{\bar{p}+\bar{c}}.$$ (15.6)

This result shows how basin runoff changes with mean rainfall for given mean soil absorption capacities under the assumption that the variation of rainfall and capacity over the basin is exponential. Figure 15.2 shows the form of this relation, and in particular how basin runoff is increased when rainfall and capacity are no longer assumed constant over the basin (when $Q = \bar{p}-\bar{c}$) but vary exponentially (when $Q = \bar{p}^2/(\bar{p}+\bar{c})$). The approach may be developed further to give a new type of rainfall-runoff model, based either on a storage capacity excess mechanism generating saturation overland flow or on an infiltration capacity excess mechanism producing Hortonian overland flow, and incorporating groundwater and channel translation com-

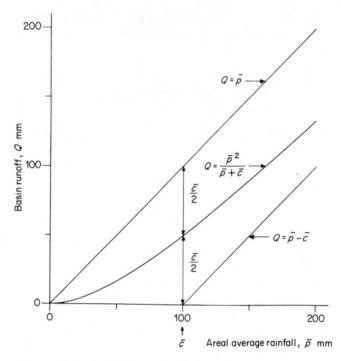

Figure 15.2. Rainfall-runoff relationship for the probability-
distributed model of rainfall and runoff production

ponents. The inclusion of rainfall variability into the above model development represents a new advance on the probability-distributed rainfall-
runoff models recently reviewed and extended by the author (Moore, 1985).

This new approach would utilize grid values of radar-derived rainfall to
identify a suitable parametric form for the density function, $f_P(p)$, and at
each time frame would use the grid data to estimate the function's par-
ameter(s). For example, the average of the grid square values would be used
to estimate the mean basin rainfall, $\bar{p}$, specifying the exponential density,
$f_P(p) = \bar{p}^{-1} \exp(-p/\bar{p})$; a new estimate of $\bar{p}$ would be obtained at each
time frame in order to calculate basin runoff. Histograms of hourly rainfall
amounts, derived from radar measurements presented on a 12×10 radar
grid (5 km square grid) located north-west of Birmingham airport and
extending over an area of 3000 km^2, are presented in Figure 15.3. These
suggest that a simple exponential density function may indeed be adequate.
A theoretical justification for rain falling on equally divided small cell regions
having an exponential distribution is given by Matsubayashi *et al.* (1984),
and empirical confirmatory evidence is provided using a dense network of
rain gauges in the vicinity of Nagoya City, Japan. As the size of the cell

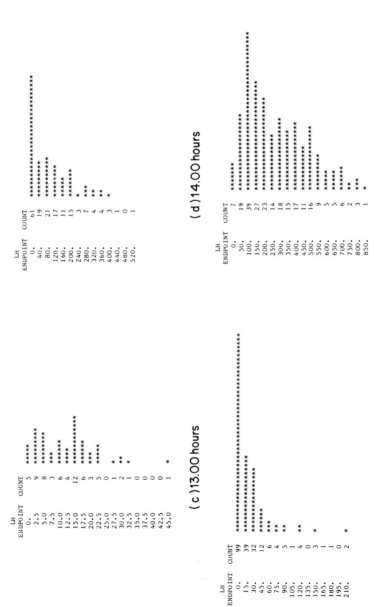

Figure 15.3. Histograms of hourly rainfall amounts (units are 0.1 mm) derived from a 12 × 10 radar grid (5 km square grid) west of Birmingham airport on 14 July 1982

increases, the distribution is shown to change to gamma and then Gaussian form.

The probability-distributed model of rainfall and soil-water capacity is seen as presenting a novel way of making greater use of spatially distributed weather radar data without resorting to the complexity of a geometrically distributed rainfall-runoff model. However, its practical utility remains to be assessed.

VALUE OF RADAR RAINFALL DATA IN CHANNEL FLOOD ROUTING MODELS

In circumstances where flood warning is to be provided at a point downstream of two or more streamflow gauging stations, then the flood forecasting model will usually consist of both channel routing and rainfall-runoff components.

Consider the problem of constructing a flow forecasting strategy for the simple river basin depicted in Figure 15.4(a). The straightforward approach

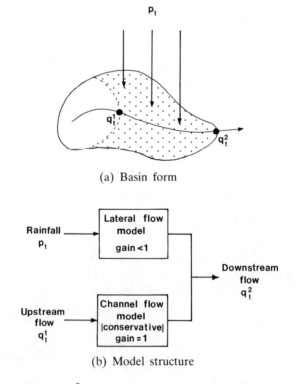

(a) Basin form

(b) Model structure

Figure 15.4. A flood forecasting model incorporating a continuity-preserving channel flow routing model and a rainfall–lateral inflow model

would be to construct a model relating the downstream flow at time t, q_t^2, to the sum of two components, one attributed to the influence of the upstream flow, q_t^1, and the other due to lateral inflow from the intermediate catchment area; Figure 15.4(b) presents a schematic of the model structure where rainfall, p_t, over the lateral inflow area is assumed to be derived from a sparse raingauge network. The component of the flow forecast deriving from the influence of the lateral inflow will be expected to account for most of the errors in the total forecast because of errors in the gauge-derived areal rainfall estimate, and the greater difficulty of modelling the rainfall-runoff process compared to modelling the channel flow routing process. Indeed, an improved forecast may result from omission of the rainfall information altogether, using instead the upstream flow, q_t^1, as input to a channel flow routing model which does not preserve continuity, but amplifies the upstream flow volume to that of the downstream flow to be forecast. This would be done in the belief that upstream flow is more representative of flow from the ungauged area contributing lateral inflow than forecasts from a rainfall-runoff model using error-corrupted rainfall measurements over the ungauged area as input.

Even if the forecast from a model which includes the raingauge data is superior in an average sense, it may not be preferred on account of its lack of 'robustness'. Here we mean that a model lacks robustness if it can give rise occasionally to very spurious forecasts on account of errors, in this case due to the use of point gauge measurements to estimate areal rainfall. In such circumstances radar data may be expected to provide more robust forecasts from rainfall-runoff models than would be obtained from a sparse raingauge network, especially in areas experiencing high spatial variability in rainfall. An argument based on lack of robustness has been used to exclude raingauge measurements from a procedure developed to forecast inflow to a reservoir located in the tropics, where rainfall is spatially highly variable and raingauges are few; forecasts were required to be robust since they were to assist in the control of a reservoir in order to mitigate flooding downstream. In such circumstances radar-derived rainfall data may be expected to provide more robust information on lateral inflow variation than can be provided by at-a-point raingauge measurements.

THE ROLE OF UPDATING METHODS IN COMPENSATING FOR RAINFALL ERRORS

The degradation of rainfall-runoff model performance by error-corrupted rainfall input data may be compensated for in real time through updating techniques which incorporate current measurements of flow at the forecast location. Updating may be accomplished by (a) error prediction, where the temporal persistence of model errors is used to predict future errors; (b) parameter adjustment; and (c) state adjustment, where corrections are made to the water content of internal storages to achieve accordance with observed flows. Reviews of these techniques are presented in Reed (1984) and Moore (1983). Error prediction may now be regarded as an established technique which is easy to implement (Moore, 1982; Jones and Moore, 1980); however, its benefit is often least in the vicinity of the hydrograph peak where model errors display least temporal persistence. Parameter adjustment, accomplished usually by techniques based on the Kalman filter (Jazwinski, 1970), appears to be a less attractive option for real-time application since parameter variation usually reflects an inadequate model structure, which can be improved by off-line studies. State adjustment is seen as a promising area for further research, and is of the most relevance to the problem of error-corrupted rainfall measurements. The effect of errors in rainfall accumulate as errors in the water contents of stores making up a conceptual rainfall-runoff model, and in turn contribute to the errors incurred in forecasting the flood hydrograph through incorrect partitioning of rainfall by the model into surface and groundwater runoff components. Conventionally, state updating is achieved through the Kalman filter, whereas this provides an exact solution for linear systems even when the random variations may be non-Gaussian; only approximate solutions may be obtained for non-linear conceptual rainfall-runoff models. There is considerable scope for developing empirical state updating schemes, which exploit the hydrologist's understanding of the physical mechanisms operating. These schemes use the model error, ϵ, to adjust the set of model store contents (the state variables): for example, the soil-water store content, S_1, and the groundwater store content, S_2. The adjustment chosen might be of the familiar Kalman filter form

$$S_i^* = S_i + K_i\epsilon \quad i = 1, 2 \tag{15.7}$$

but the gains K_1 and K_2 would be derived empirically through optimization. Dependence of K_1 and K_2 on measured discharge or on the model flow components could also be incorporated in physically sensible ways into this empirical state updating scheme. Whilst not removing the need for accurate rainfall measurements up to the present, and forecasts of rainfall in the future, state adjustment provides a mechanism for compensating for the effect of past error-corrupted rainfalls on the internal states of the model which affect the important separation of rainfall into its storm and baseflow runoff components.

THE HYDROLOGICAL USE OF RADAR-DERIVED FORECASTS OF RAINFALL

Perhaps the most serious shortcoming of rainfall-runoff models for flood forecasting is their dependence on the availability of rainfall forecasts in order to forecast flows at lead times beyond the natural time lag of the catchment. Radar's ability to portray the pattern of rainfall over space, and to allow the movement and the development of the pattern to be monitored over time, give rise to high expectations of radar's value for rainfall forecasting. Methods which have been developed for rainfall forecasting using weather radar are largely based on displacement of the radar-derived grid square estimates so as to maximize some measure of association with values obtained at subsequent time frames; examples are provided by Austin and Bellon (1974) in the USA, by Hill *et al.* (1977) in the UK, and Yoshino (1985) in Japan. The hydrologist still awaits refinements to these techniques, studies evaluating their worth for flood warning, and their provision in real time for use in operational flood warning and reservoir control schemes.

EVALUATING RADAR-DERIVED RAINFALL FOR FLOOD FORECASTING

Having discussed some ways in which weather radar might be more effectively used for flood forecasting, a re-evaluation of the benefit of radar relative to conventional raingauge networks is called for, which incorporates these ideas. An evaluation framework for assessing the value of radar-derived rainfall data for flood forecasting may take two forms. The indirect approach is to assess the weather radar's ability to estimate areal average rainfall, a quantity often stated as a requirement for data input to lumped rainfall-runoff models. An assessment of this type is problematic due to the lack of 'truth data'. The expensive solution of comparing radar estimates of areal rainfall with estimates derived from a special dense network of recording raingauges has been used in the past, for example as part of the Dee Weather Radar Project (Central Water Planning Unit, 1977; Harrold *et al.*, 1974). A more satisfactory evaluation in terms of radar's value specifically for flood forecasting would be based on a direct assessment of the forecasting performance it provides relative to that obtained from the use of data from recording raingauge networks of differing density and configuration. Given the earlier discussion of the effect of rainfall input errors on rainfall-runoff model performance, it would clearly be necessary to calibrate models using both sources of data in order to achieve a fair comparison. Special consideration would also need to be given to the rainfall-runoff model structures to be used, since these would influence the final outcome. In particular, models which fully exploit the spatial nature of radar-derived rainfall data should be included. To achieve a fair comparison with data derived from a raingauge network, techniques of spatial interpolation using rain gauge data would need to be employed in order to provide data input to grid square distributed models used in the evaluation (Anderl *et al.*, 1976; Bastin *et al.*, 1984). For the results of the evaluation to have any credibility a reasonably large number of flood events would be needed for calibration, together with a further set of events for use in an independent model evaluation. The evaluation should also be carried out for a range of catchments of differing hydrological character, in order to construct guidelines for weather radar's utility across a range of hydrological situations. In particular, the intense and localized nature of rainfall in semi-arid regions of the world provides a strong *a priori* argument in favour of weather radar, as opposed to gauge-based, rainfall measuring networks for use in flood forecasting, but this has yet to be demonstrated in practice. At present no rigorous direct evaluation across a range of hydrological environments has been made, and indeed the database to support such a comparison is still not available. Such an evaluation is clearly a crucial requirement for the future.

CONCLUSION

Collier (1984) provides a cogent summary of the present state of the art in the use of radar rainfall data for flood forecasting. He states that

> it is still not clear whether the accuracy quoted in the literature can be reproduced consistently in a fully operational system, or indeed whether that accuracy is acceptable for real-time flood forecasting. There have been few, if any, reports describing radar systems which make quantitative measurements continuously (24 hours per day) over a long period during which data have been supplied direct to hydrological and meteorological users.

This points to a future requirement for a more extensive evaluation of flood forecasting performance based on radar rainfall data and the need for continuous intra-storm data sets on which to base such an appraisal. However there is also a requirement for the hydrologist to develop real-time flood forecasting techniques which make full use of the spatially distributed picture of rainfall that radar provides, and which compensate for its variable precision. The present paper has made some suggestions as to how this development might proceed.

ACKNOWLEDGEMENTS

This contribution was prepared using funding from the Ministry of Agriculture, Fisheries and Food under their Flood Protection Commission programme.

REFERENCES

Abbott, M. B., Clarke, R. T., and Preissman, A. (1978). Logistics and benefits of the European Hydrologic System. *Proc. Int. Symp. on Logistics and Benefits of using Mathematical Models of Hydrologic and Water Resource Systems*, Pisa, Italy.

Anderl, B., Attmannspacher, W., and Schultz, G. A. (1976). Accuracy of reservoir inflow forecasts based on radar rainfall measurements. *Water Resources Research*, **12**(2), 217–23.

Austin, G. L., and Bellon, A. (1974). The use of digital weather radar records for short-term precipitation forecasting. *Quart. J. R. Met. Soc.*, **100**, 658–64.

Bastin, G., Lorent, B., Duque, C., and Gevers, M. (1984). Optimal estimation of the average areal rainfall and optimal selection of raingauge locations. *Water Resources Research*, **20**(4), 463–70.

Central Water Planning Unit (1977). Dee weather radar and real-time hydrological forecasting project. Report by the Steering Committee, 172 pp.

Collier, C. G. (1984). The operational performance in estimating surface rainfall of a raingauge-calibrated radar system. 22nd Conf. on Radar Meteorology, Zurich, Switzerland. *Am. Met. Soc., Boston, Mass. USA*, pp. 257–62.

Collier, C. G., Rouke, P. R., and May, B. R. (1983). A weather radar correction procedure for real-time estimation of surface rainfall. *Quart. J. R. Met. Soc.*, **109**, 589–608.

Hamlin, M. J. (1983). The significance of rainfall in the study of hydrological processes at basin scale. *J. Hydrology*, **65**(1/3), 73–94.

Harrold, T. W., English, E. J., and Nicholass, C. A. (1974). The accuracy of radar-derived rainfall measurements in hilly terrain. *Quart. J. R. Met. Soc.*, **100**, 331–50.

Hill, F. F., Whyte, K. W., and Browning, K. A. (1977). The contribution of a weather radar network to forecasting frontal precipitation: a case study. *Meteorol. Mag.* **106**(1256), 69–89.

Jazwinski, A. H. (1970). *Stochastic Processes and Filtering Theory*. Academic Press, 376 pp.

Jones, D. A., and Moore, R. J. (1980). A simple channel flow routing model for real time use, *Hydrological Forecasting, Proc. Oxford Symp.*, IAHS-AISH publ. no. 129, pp. 397–408.

Matsubayashi, O., Takagi, F., and Tonomura, A. (1984). The probability density function of areal average rainfall. *J. Hydrosci. Hydraulic Eng* **2**(1), 63–71.

Moore, R. J. (1977). Raingauge network requirements for real-time flow forecasting. In O'Connell, P. E., Beran, M. A., Gurney, R. J., Jones, D. A., and Moore, R. J. Methods for evaluating the UK raingauge network. Institute of Hydrology Report No. 40, pp. 168–182, 190–205.

Moore, R. J. (1982). Transfer functions, noise predictors and the forecasting of flood events in real-time. In Singh, V. P. (ed.), *Statistical Analysis of Rainfall and Runoff*, Water Resources Publications, Colorado, pp. 229–50.

Moore, R. J. (1983). Flood forecasting techniques. WMO/UNDP Regional Seminar on Flood Forecasting, Bangkok, Thailand, 37 pp.

Moore, R. J. (1985). The probability-distributed principle and runoff production at point and basin scales. *Hydrol. Sc. J.* **30**(2), 273–97.

Morris, E. M. (1980). Forecasting flood flows in grassy and forested basins using a deterministic distributed mathematical model. *Hydrological Forecasting. Proc. Oxford Symp.*, IAHS-AISH publ. no. 129, pp. 247–55.

Ngirane-Katashaya, G. G., and Wheater, H. S. (1985). Hydrograph sensitivity to storm kinematics. *Water Resources Research*, **21**(3), 337–45.

Niemczynowicz, J. (1984). An investigation of the areal and dynamic properties of rainfall and its influence on runoff generating processes. Report no. 1005, Dept. of Water Resources Engineering, Lunds Institute of Technology/University of Lund, 215 pp.

Reed, D. W. (1984). A review of British flood forecasting practice. *Institute of Hydrology Report No. 90*, 113 pp.

238 *Weather radar and flood forecasting*

Singh, V. P. (1977). Sensitivity of some runoff models to errors in rainfall excess. *J. Hydrol.* **33**, 301–318.

Singh, V. P., and Woolhiser, D. A. (1976). Sensitivity of linear and nonlinear surface runoff models to input errors. *J. Hydrol.* **29**, 243–249.

Troutman, B. M. (1982). An analysis of input errors in precipitation-runoff models using regression with errors in the independent variables. *Water Resources Research* **18**(4), 947–64.

Troutman, B. M. (1983). Runoff prediction errors and bias in parameter estimation induced by spatial variability of precipitation. *Water Resources Research* **19**(3), 791–810.

Yoshino, F. (1985). Study on short-term forecasting of rainfall using radar raingauge. Public Works Research Institute, Ministry of Construction, Japan, 20 pp.

Weather Radar and Flood Forecasting
Edited by V.K. Collinge and C. Kirby
© 1987 John Wiley & Sons Ltd.

CHAPTER 16

Towards the More Effective Use of Radar and Satellite Imagery in Weather Forecasting

K. A. Browning

INTRODUCTION

This chapter is based on a paper presented at two meetings in 1985; although these meetings were concerned primarily with the use of weather radar, the cloud imagery obtainable from a geostationary satellite is so complementary to the precipitation patterns from ground-based radars, and the challenge of using them effectively is so similar, that the scope of the original paper has been extended to cover the overall issue of how to exploit both kinds of imagery.

The two meetings in 1985 mark milestones in the development of weather radar and related technology in Europe. The first meeting, at the University of Lancaster (UK) on 16–18 September, dealt with the use of weather radar for flood forecasting. It commemorated the success of the unmanned operational radar installation at Hameldon Hill, part of the UK weather radar network that became operational in 1985. The second meeting, at Erice (Sicily) on 30 September to 4 October, commemorated the work of the COST-72 committee towards establishing the feasibility of a European network of weather radars. The central idea behind COST-72 is that by combining digital data from a network of radars it is possible to observe precipitation systems in their entirety, and to distribute the resulting composite pictures widely for flexible use by many users. As a result of the work of COST-72, and of individual nations, it now seems likely that weather radar, supplemented by the European geostationary satellite, Meteosat,

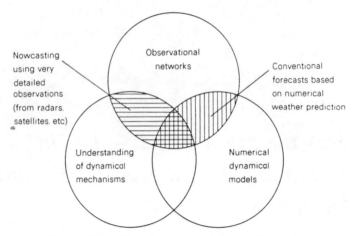

Figure 16.1. The observing and forecasting system

will play an increasingly important role in meteorological and hydrological forecasting over the coming decade (see Chapter 2 for more details).

This chapter concentrates on meteorological forecasting. To understand in the broadest sense the role of imagery from radar networks and Meteosat, it is necessary to keep in mind the unique characteristics of the imagery as a source of meteorological information. On the positive side it combines wide coverage of the patterns of precipitation and cloud with high resolution both in space and time. To be set against this, however, is the fact that it is not easy to use the data quantitatively, especially within the context of numerical weather prediction. Thus two rather separate weather forecasting procedures have grown up side by side, as shown in Figure 16.1. On the one hand we have numerical–dynamical models, using quantitative data from conventional *in situ* observations plus satellite soundings, but not yet using much of the information from satellite or radar imagery. This approach is particularly successful for forecast periods of a day or more ahead. On the other hand, we have so-called nowcasting methods in which the detailed descriptions of the current weather, especially the distribution of precipitation and cloud from radar and satellite, are used to identify mesoscale weather phenomena and to provide highly specific forecasts by linear extrapolation for very short periods ahead. The period of valid extrapolation is six or more hours in the case of some frontal systems but may be limited to an hour or less for individual thunderstorms.

The reasons for simple extrapolation breaking down after so short a time are the development and decay of rain and cloud areas, and changes in the velocity of these areas. The new developments are forced by the topography or by synoptic-scale dynamical systems and, as will be explained later, there is some scope for extending the forecast period if we can understand what is happening dynamically. Conceptual models of the structure, mechanism and life cycle of various kinds of weather system will thus be useful for providing what might be called extended nowcasts. Certain aspects of the detailed imagery may also be useful for initializing numerical–dynamical models. This is not a straightforward task, however, since radar and satellite images do not directly represent any of the dynamical variables (temperature, pressure, wind, humidity) needed as input to numerical models. It will thus be necessary to develop indirect methods of incorporating information from the imagery into the models.

The marriage of detailed imagery from radar and satellite with conceptual and numerical weather prediction models constitutes one of the major challenges for meteorologists during the coming decade. The challenge spreads across the areas of dynamical meteorology, numerical techniques, and computer and graphics display technology.

METHODS OF USING RADAR AND SATELLITE IMAGERY

The precipitation and cloud imagery from radar networks and from satellites can be exploited in several ways:

1. to provide measurements of precipitation amounts and of cloud height and cover;
2. to identify specific categories of weather phenomena, especially on the mesoscale;
3. to provide very-short-range forecasts by simple extrapolation, i.e. 'nowcasting';
4. to provide extended nowcasts benefiting from life cycle conceptual models of weather phenomena;
5. to help in the adjustment of the output of a numerical dynamical model; and
6. to initialize numerical dynamical prediction models, especially mesoscale models.

Measurement of precipitation

The use of radar to measure precipitation intensity, and satellite imagery to measure cloud height and coverage, are both well-established techniques and will be mentioned only briefly here. In the case of the measurement of precipitation by radar there are various limitations in accuracy brought

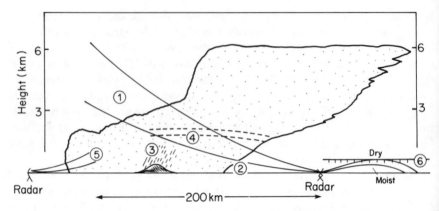

Figure 16.2.   Cross-section through an area of frontal precipitation illustrating six sources of error in the radar measurement of surface rainfall intensity, namely: (1) radar beam overshooting the shallow precipitation at long ranges, (2) low-level evaporation beneath the radar beam, (3) orographic enhancement above hills which goes undetected beneath the radar beam, (4) anomalously high radar signal from melting snow (the bright band), (5) underestimation of the intensity of drizzle because of the absence of large droplets, and (6) radar beam bent in the presence of a strong hydrolapse, causing it to intercept land or sea

about more by meteorological factors (Figure 16.2) than by instrumental measurement errors. Although good accuracies have been achieved in some experimental studies, in operational practice it is prudent not to rely on accuracies much better than plus or minus a factor of two unless expensive precautions can be taken, such as operating only at close range or having large numbers of telemetering rain gauges for 'calibration' purposes.

Satellite cloud imagery can be used to infer the precipitation distribution in areas where radar and other measurements are lacking. A useful approach is to use bispectral information from Meteosat. Infra-red imagery provides an indication of cloud top height, and visible imagery an indication of cloud thickness. Together they provide a better indication of probable precipitation than either can alone. At night one must depend mainly on the infra-red channel, although the Meteosat water vapour channel may add a little information.

The transfer function between the observed cloud properties and surface precipitation is rather variable and so, although the satellite information provides a useful indication of precipitation, it is still necessary to use radar whenever possible. In the example shown in Plate 7 this is achieved by using radar-derived measurements to the exclusion of satellite data where radars exist over the UK (within the white outline) whilst using satellite data beyond radar range. It is of course important to reduce both sets of data to a common format which, in this case, is a transverse Mercator projection with a 5 × 5 km cell size. The two sets of data are seen to fit together quite well in Plate 7. This is often true for showers and cold frontal belts of rain; however, the relationship between cloud and precipitation tends to be poor where, as at some warm fronts, the region of upper cloud overruns a dry layer and hence extends far ahead of the surface precipitation (as in Figure 16.2).

The satellite data beyond radar range in Plate 7 were 'calibrated' in terms of probable surface precipitation using a relationship derived by comparing current satellite and radar data within the area of combined coverage. Obviously the quality of such calibration is likely to diminish with distance from the region where the calibration was derived, especially where there is a change in synoptic type. When a precipitation system is situated entirely beyond radar coverage or there is a different synoptic regime outside the radar area, then it becomes necessary to use climatological transfer functions. Much work remains to be done to develop an appropriate set of transfer functions for different situations. Given the set of transfer functions it will be necessary to provide facilities which enable different relationships to be applied to different parts of the cloud distribution as selected by a meteorologist.

Identification of mesoscale weather phenomena

For a forecaster in a regional weather office, one of the principal benefits offered by radar and satellite imagery is the ability to identify phenomena that are too small to be resolved clearly by conventional observations. Characteristic patterns in the distribution of precipitation and cloud indicate the presence of phenomena such as sharp cold fronts, subsynoptic positive vorticity advection centres, and mesoscale convective systems. Given an understanding of these phenomena crystallized in the form of conceptual models, the forecaster is able to make a much finer scale analysis than would otherwise have been possible. Indeed, by interpreting his conventional observations and model products within the context of the imagery, he is able to sharpen up his analysis not only of precipitation and cloud but also of other variables which are known to bear a close relationship to the precipitation and cloud.

An example is shown in Plate 8, in which the narrow band of heavy rain seen by radar is associated with line convection. Line convection is a well-defined mesoscale phenomenon occurring at sharp cold fronts in which a cold density current at the surface forces a narrow band of intense convection ahead of it in the bottom 3 km of the atmosphere. A conceptual model representing the flow in a section normal to the line is shown in Figure 16.3. The line convection in Plate 8 was travelling eastwards towards London. Observations at stations over which it passed showed abrupt and charac-

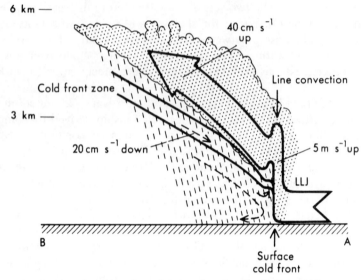

Figure 16.3.   Conceptual model showing the circulation within a vertical section normal to a sharp cold front

teristic changes in wind, temperature, pressure, visibility and cloud base. These characteristic changes accompanied the line convection throughout its passage across central England. Given this information it would thus have come as no surprise when, shortly after the time of Plate 8, the line convection passed over London Heathrow with the sequence of events shown in Figure 16.4.

Line convection is a synoptically forced phenomenon associated with a certain kind of cold front. Topographically forced phenomena are also well revealed by radar, although much remains to be done to clarify and explain the variety of topographical effects that can occur in regions of complex topography. An example of one of the more clear-cut topographically forced phenomena is given in Figure 16.5. It shows a band of showers extending south-eastwards from the north Irish Sea. This is a common event during cold north-westerly outbreaks in winter. Observations of a larger area by satellite show that the showers come from the stretch of sea between Ireland and Scotland called the North Channel. North Channel shower bands often persist for long periods, occasionally up to a day or more. Series of showers, with poor visibility and perhaps snow, can then affect places such as Manchester Airport hour after hour, at a time when most parts of the country are enjoying relatively good weather. It is clearly important to be able to identify this kind of local anomaly, and radar and satellite imagery provide the best means of doing so.

Satellite imagery is very helpful for identifying and categorizing deep

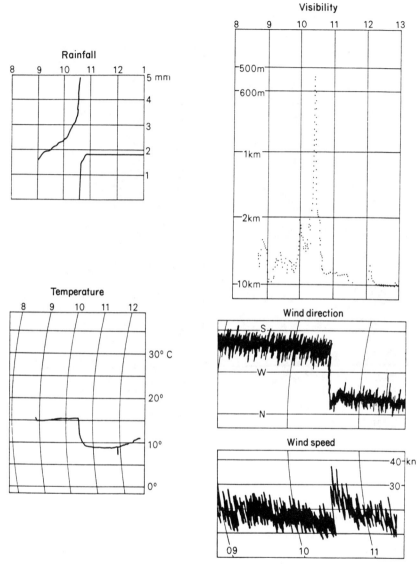

Figure 16.4. Autographic charts showing the abrupt changes in the weather at London Heathrow during the passage of the line convection depicted in Plate 8

convective phenomena. This includes the topographically forced patterns of convection discussed above. It also includes individual tall cumulonimbus clouds which, when they extend up to or above the level of the tropopause, are liable to produce severe weather. Another important category of phenomenon well observed by satellite is the Mesoscale Convective Complex

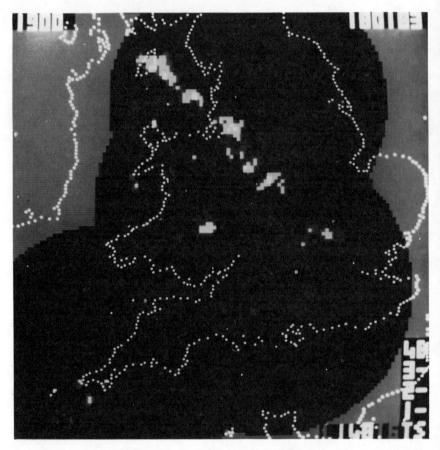

Figure 16.5.   Mesoscale band of showers extending into central England
downwind of the North Channel gap between Ireland and Scotland, as
observed by the UK weather radar network

or Mesoscale Convective System (MCS). As in the example in Plate 10 an
MCS appears as a cold and rather symmetrical cirrus cloud shield generated
from the combined anvil outflow from a cluster of deep thunderstorm clouds.
The radar network gives a useful complementary view and in Plate 11 the
intense convective precipitation can be seen in the southern parts of the
MCS, with an area of widespread uniform rain to the north of that. The
satellite and radar data can be interpreted within the context of the conceptual model in Figure 16.6, which shows that deep convection is triggered
by air with high wet-bulb potential temperature encountering a barrier of
cold air. The cold air barrier is associated partly with a pre-existing front
but is also enhanced by evaporative cooling. The deep convection occurs on

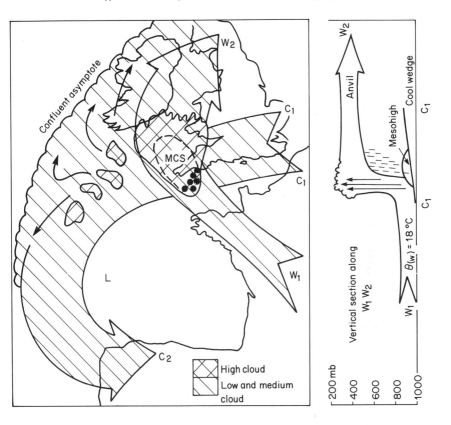

Figure 16.6. Conceptual model of the mesoscale convective system portrayed in Plates 10 and 11

the upstream (southern) side of the MCS with more gentle slantwise ascent on the downstream (northern) side.

So far, where satellites are concerned, we have concentrated on the cloud imagery. However, the water vapour (WV) channel on Meteosat also provides useful information about mesoscale weather systems in some situations. Thus for example Figure 16.7 shows the WV imagery on an occasion (24 April 1984) when a cold pool lay over the English Channel. Cold pools are often associated with outbreaks of thunderstorms (e.g. the MCS in Plate 10), but the cold pool on 24 April was different in that it occurred within the circulation of an anticyclone in a region of mainly clear skies. Although the cold pool was restricted to the layer 700 to 400 mb and was only 500 km in diameter, its circulation was distinct enough to perturb

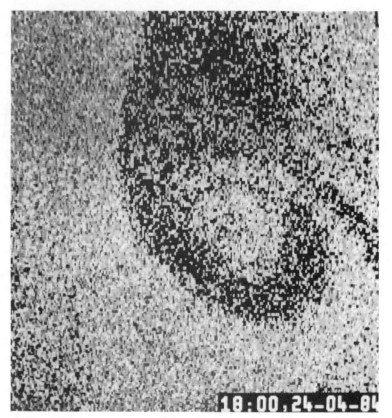

Figure 16.7. Circulation around a cold pool as seen in Meteosat water vapour imagery. Black and white represents respectively dry and moist air in the upper troposphere

the pattern of humidity in the upper troposphere and produce the pronounced spiral shape seen in Figure 16.7. The infra-red imagery in Figure 16.8 shows a few shallow convective showers over the English Channel, close to the centre of the spiral. From time to time these showers developed into thunderstorms but the storms behaved in a rather erratic manner. The spiral pattern in the WV imagery, on the other hand, could be tracked as a major eye-catching feature for more than a day whilst showers and thunderstorms seen by radar and satellite infra-red imagery developed, decayed and subsequently redeveloped at its centre. In isolation the WV imagery is difficult to use but it can be valuable when taken together with other data.

Extrapolation forecasting

Slow-moving 'air mass' thunderstorms are notoriously erratic in their behaviour. It is sometimes possible to obtain a useful forecast by extrapolating the

Figure 16.8. Infra-red imagery from Meteosat at a time corresponding to Figure 16.7 showing convective showers over the channel between France and south-west England located at the centre of the cold pool. The showers are shown mainly dark grey. Elsewhere there were extensive clear skies with cold seas appearing white and warm land blank

envelope of a cluster of such storms, but the individual storm cells tend to be short-lived, with new ones forming rather unpredictably within the envelope. In other circumstances, however, especially for severe storms, even individual cells may have sufficient persistence to be amenable to extrapolating forecasting. This was the case for the fast-moving severe local storms shown in Figure 16.9. The storm that travelled along the south coast of Cornwall had been moving at 90 km h^{-1} for 3 h before it lashed South Devon with large hail.

The convective showers that approach NW England in North Channel shower band situations of the kind described above (Figure 16.5) are of only moderate intensity; yet they, too, tend to be long lived. Figure 16.10 shows the successive $\frac{1}{2}$-hourly positions of four of the shower clouds depicted in Figure 16.5. It can be seen that the showers that penetrated inland over

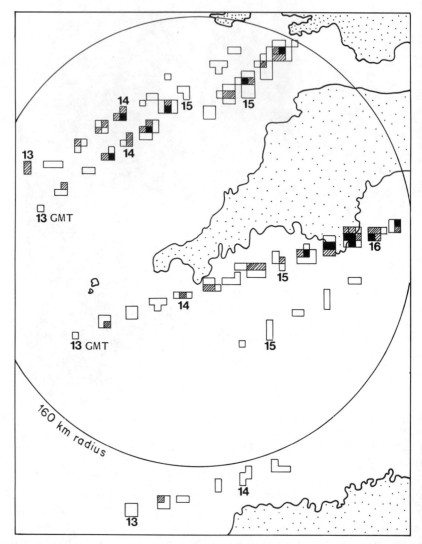

Figure 16.9. Successive positions of the cores of several severe storms at 15-min intervals as seen by a radar in south-west England. (Courtesy of R. G. Owens.)

central England were either pre-existing showers that travelled through the North Channel gap or else they formed within the North Channel gap at a time when a nearby shower dissipated on encountering high land flanking the North Channel. In either event, however, the individual showers that reached England all had lifetimes in excess of 4 h.

Frontal cloud systems are also amenable to extrapolation forecasting, although the forecaster must watch out for a number of pitfalls. One problem

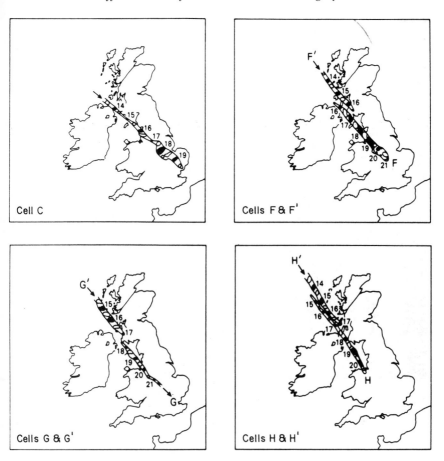

Figure 16.10. Successive positions of several shower clouds at 30-min intervals as seen in Meteosat infra-red imagery. Note the persistence of shower clouds travelling through or forming within the North Channel gap between Ireland and Scotland

is differential motion of areas of cloud and rain, especially in the vicinity of a circulation centre. Another problem is that mesoscale precipitation areas within frontal rainbands tend to travel faster than the synoptic scale frontal system with which they are associated. Thus, mature mesoscale precipitation areas tend to decay at the leading edge of the frontal system while new ones form near the rear edge. Despite these problems it is found that quite small areas of precipitation, such as the cores of heavy rain shown as white in Figure 16.11, can be tracked travelling steadily for five or more hours and over several hundreds of kilometres as shown in Figure 16.12. These precipitation features are associated with mid-level convective cells which travel rapidly with the winds in the middle troposphere. The lower tro-

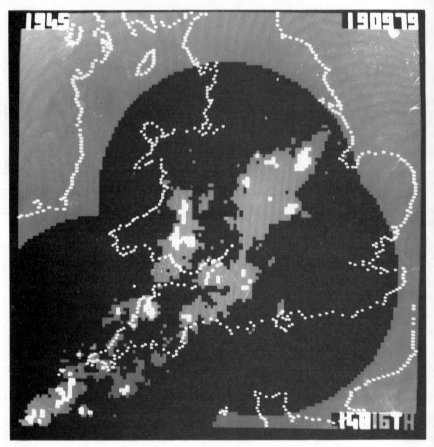

Figure 16.11.   Bands of mesoscale precipitation areas associated with a cold front. Areas of light and moderate rain are shown grey. Cores of heavier rain are shown white

pospheric line convection at cold fronts, referred to earlier, also tends to travel in a well-behaved fashion which makes it amenable to simple extrapolation forecasting. In the example shown in Figure 16.13 the line convection travelled uniformly for over 10 h. Despite the shallowness of the phenomenon, its uniformity of motion does not appear to be affected by passage over land several hundred metres high any more than was the case for the mid-level convective areas in Figures 16.11 and 16.12.

Extrapolation forecasting is conceptually very simple. However, skill has to be exercised in two respects. First, the forecaster has to make allowances for inadequacies in the imagery (e.g. range dependency in the radar pictures and uncertain cloud/rain transfer functions in the case of the satellite images). This topic is discussed further in a later section. Second, the forecaster has to

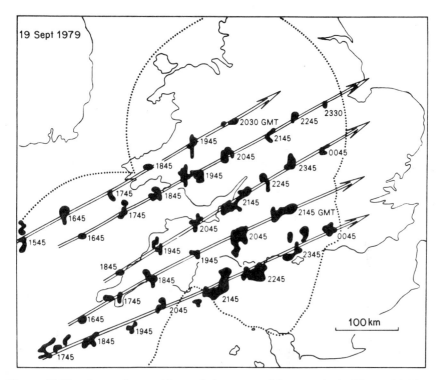

Figure 16.12. Successive positions of the cores of heavy rain in Figure 16.11 at 1 h intervals during the passage of the cold front across England and Wales

assess the range of valid extrapolation. This is very phenomenon-dependent. Sometimes it is possible to make use of conceptual life cycle models, a topic discussed in the next section.

Use of life cycle models

Forecasts produced by extrapolating mesoscale features are vulnerable to various changes in the weather system being tracked. In the case of topographically forced phenomena, such as the North Channel showers discussed earlier, changes in the phenomenon may be anticipated when there is a change in wind direction or air mass stability. In the case of mesoscale phenomena that are forced by a larger-scale dynamical system, the mesoscale phenomenon goes through a life cycle as it moves from a region of large-scale ascent to one of large-scale descent. This applies, for example, to mesoscale convective system (MCSs) of the kind shown in Plates 10 and 11 and Figure 16.6, and also to the smaller MCSs, characterized by localized regions of high clouds, that are sometimes found within subsynoptic comma cloud systems (Plate 9). The sequence of sketches in Figure 16.14 shows

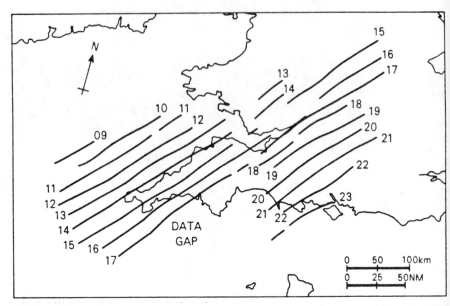

Figure 16.13. Successive hourly positions of line convection as seen by the UK weather radar network during the passage of a sharp cold front across southern England

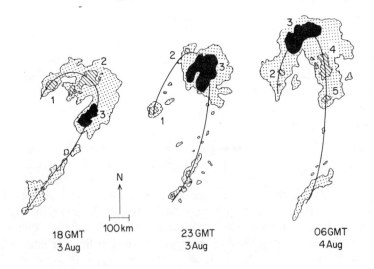

Figure 16.14. Sketches of the infra-red Meteosat cloud imagery at three times showing areas of high cloud associated with five mesoscale convective systems travelling along the axis of the comma cloud portrayed in Plate 9. (Courtesy R. Brown.)

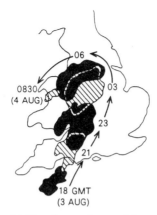

Figure 16.15. Successive positions of the region of high cloud associated with the mesoscale convective system labelled no. 3 in Figure 16.14. (Courtesy R. Brown.)

several compact regions of high cloud associated with small MCSs moving around the axis of the comma cloud system as it travelled towards the north-east. Figure 16.15 shows the life history of one of these MCSs; in terms of the extent of high cloud there is seen to have been a period of development up to 2300 GMT as the MCS travelled northwards along the tail of the comma cloud system, then a quasi-steady-state phase, followed by abrupt decay after 0500 GMT as the MCS began to curve back towards the south around the northernmost parts of the comma. There was a good relationship between high cloud tops and surface rain and, as shown by Figure 16.16, the life history was similar when expressed in terms of the extent of significant surface rain. Other MCSs have been observed to behave similarly as they travel around the axis of a comma cloud.

In the absence of a distinct front or a subsynoptic comma cloud feature, the development of thunderstorms over land may appear to be a chaotic process when viewed by radar, yet when viewed by satellite new thunder-storms are sometimes seen to form systematically on thin arc clouds demar-cating the cold air outflows from previous storms. The development of new storms is particularly favoured where two such arc clouds intersect. This phenomenon is familiar in the United States but less so in Europe. This is partly because thunderstorms producing large cold air outflows at the surface are less common in Europe, and partly perhaps because the lines of cumulus forming the arc clouds are more easily detected in the otherwise less cloudy conditions of the USA. Their detection is also helped in the USA by the higher spatial resolution available in the visible GOES imagery than in Meteosat.

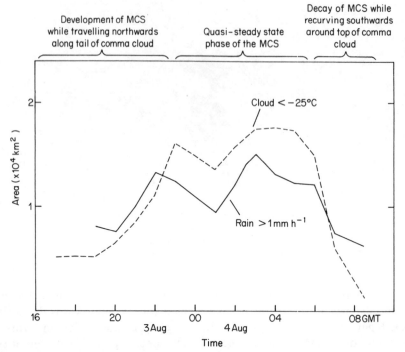

Figure 16.16. Temporal variation of the extent of cold high cloud and moderate rain associated with the mesoscale convective system labelled no. 3 in Figure 16.14. (Courtesy R. Brown.)

Simple conceptual life cycle models need to be developed for a variety of situations. When applied to the frequent imagery from radar and Meteosat they help a forecaster anticipate when his simple extrapolation forecasts are liable to go astray; however, the application of appropriate models in the hybrid situations often encountered in the real world will require considerable skill.

Interpretation and possible adjustment of numerical model output in the light of imagery

Radar and satellite imagery often provide some of the earliest indications that numerical predictions may be coming, or failing to come, to fruition. Sometimes the imagery will reveal a feature predicted by the model, thereby giving confidence in the model output although perhaps indicating that the feature is predicted in the wrong place or at the wrong time. Provided the model predictions are broadly credible and the forecaster has a conceptual model that enables him to interpret the imagery in terms of the variables

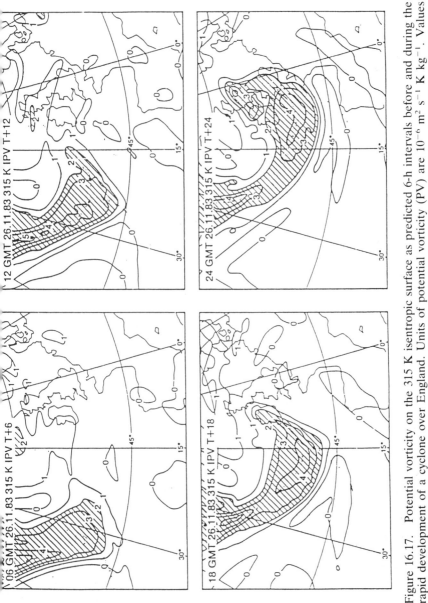

Figure 16.17. Potential vorticity on the 315 K isentropic surface as predicted 6-h intervals before and during the rapid development of a cyclone over England. Units of potential vorticity (PV) are 10^{-6} m^2 s^{-1} K kg^{-1}. Values more (less) than 2 PV units are associated with air of stratospheric (tropospheric) origin. The hatched area corresponds to an intrusion of a tongue of stratospheric air downwards into the troposphere

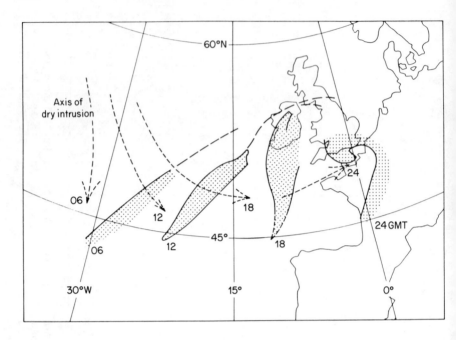

Figure 16.18.   Successive 6-hourly positions of the axis of an intrusion of dry air (dashed lines), corresponding to the tongue of high potential vorticity in Figure 16.17, with the associated baroclinic leaf cloud detected by satellite shown stippled

predicted by the model, it may be possible for him to adjust the model predictions in space or time to fit the observed imagery as it unfolds.

An example of a good fit between model predictions and satellite observations is given in Figures 16.17, 16.18 and Plate 12. This was an occasion when rapid cyclogenesis was expected as an upper tropospheric jet streak travelled around a major trough, and air with high potential vorticity descended and overran the baroclinic zone ahead of the trough. Figure 16.17 shows predictions of potential vorticity at four successive times as given by the Meteorological Office operational fine-mesh model. The region of high potential vorticity corresponded to an intrusion of dry air descending from the lower stratosphere. The axis of this dry intrusion is plotted in Figure 16.18 for the four times depicted in Figure 16.17. Ahead of the dry intrusion there was a sharp gradient of humidity and, indeed, Figure 16.18 shows sketches of a well-defined feature referred to as a baroclinic cloud leaf, which was detected in the satellite imagery and which corresponded well with the model-derived moist region at the leading edge of the dry intrusion.

The cloud leaf can be seen to have rotated as the dry intrusion overtook it. The greatest rotation occurred at the time of rapid cyclogenesis. Plate 12 is a Meteosat water vapour image which shows the dry intrusion in the upper troposphere after it had wrapped itself around the southern tip of the leaf cloud. The good relationship in this case between the features observed by satellite and the model predictions would have enabled a forecaster to have had greater confidence in the subsequent evolution as predicted by the model.

Using imagery as one of the inputs for initializing numerical dynamical models

Operational numerical prediction models, even those described as fine mesh, are too coarse to provide detailed area-specific forecasts. They are best suited for providing general forecasts for one or more days ahead. At the other extreme, simple extrapolation forecasts using radar and satellite imagery tend to break down for forecasts beyond a few hours. Mesoscale numerical models, using a grid scale of order 10 km, are under development at the Meteorological Office to close the gap between these two approaches. In situations when the predominant forcing is by topographical features such as land–water boundaries and hills, a mesoscale model can in principle be expected to perform usefully even in the absence of detailed observational inputs. However, in situations with strong synoptic forcing it is crucial to represent that forcing in the initial conditions. A general background description can be provided by synoptic scale models but further detail has also to be provided. Existing *in-situ* observations and satellite soundings are inadequate by themselves, and it is necessary to find ways of using radar and satellite imagery despite the fact that neither form of imagery represents directly the required dynamical variables. This will call for the development over the next decade of a variety of indirect approaches, some of which will now be considered in outline.

Introducing a more realistic distribution of vertical velocity

Satellite and radar imagery provides a qualitative indication of the sign of vertical motion when the images are interpreted within the context of conceptual models. Thus, for example, mesoscale regions of ascent occur within subsynoptic comma clouds and are followed by mesoscale regions of descent which give rise to the relatively cloud-free zone behind them. The vertical velocity in the numerical model is best adjusted via the field of horizontal divergence. Imbalances between the wind and mass fields are established, and can be maintained provided a diabatic heat source is available. This leads on to the next topic.

Introducing more realistic latent heat effects through improved humidity analyses

A knowledge of the detailed distribution of cloud and precipitation from satellite and radar imagery, supplemented by surface observations of cloud base etc., can enable the three-dimensional distribution of relative humidity and cloud water content to be estimated. If such an analysis yields high humidities in a region where the mesoscale model is already inclined to generate rising motion, or where rising motion is imposed by the analyst via the divergence field, the effect of latent heat release is to generate positive feedback, thereby intensifying the vertical motion. Realistic enhancement of mesoscale circulations can be achieved only if the background field provided by the synoptic scale model is reasonably accurate. In the event of a mismatch between the vertical velocities produced by the background field and the regions of high relative humidity implied by the imagery, any such area of high humidity imposed on the model would quickly decay.

Sharpening gradients

Imagery sometimes provides an indication of strong frontal discontinuities. The true gradient is available from *in situ* observations but the imagery allows the pattern to be interpolated between point observations. Thus, for example, as discussed earlier, radar can detect the narrow rainband associated with line convection. Hence the position of a sharp surface cold front can be located precisely. The mesoscale model under development at the Meteorological Office receives a broad indication of frontal features as a background field from a synoptic scale model, but such a field will inevitably be highly smoothed. Provided the cold frontal position given by the background field is broadly correct, it should be possible to sharpen up the frontal structure in the mesoscale model without upsetting the balance between wind and mass fields. This can be expected to lead in turn to a sharpening of the circulation and hence the intensification of the model-derived rainband. It is important, however, to ensure that the horizontal diffusion in the model is not too great; otherwise the impact of any imposed sharpening will be quickly damped out.

Introducing a realistic distribution of subgridscale convection

A common problem with synoptic scale numerical models is the under-prediction of rain, for example, at some cold fronts. This may happen when the large-scale vertical velocity is weak and much of the saturated ascent is taking place on smaller scales in association with vertical convection. An analysis of the texture of the satellite and radar imagery can be used to

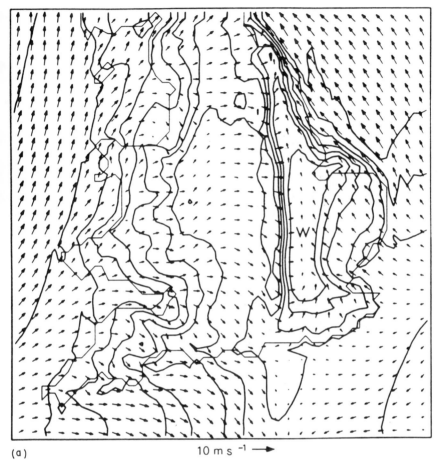

(a)   10 m s $^{-1}$ ⟶

Figure 16.19.   Mesoscale model predictions of surface wind and surface temperature (solid isopleths at intervals of 1 K) (a) with, and (b) without, the imposition of a sharp cloud edge. (Courtesy of Dr. K. M. Carpenter.)

identify the occurrence of small-scale convection. Then, given information on the cloud base and cloud top height, the effect of the convection can be parameterized. The convection leads to non-linear enhancement of the ascent through the release of latent heat.

Introducing the effects of cloud boundaries on the radiation balance

Cloud cover has a major effect over land in decreasing the temperature of the underlying ground and of the air close to it. Thus the presence of a well-defined boundary to a cloud sheet can lead to the development of solenoidal circulations, resembling inland sea breezes. Figures 16.19 (a and b) reveal

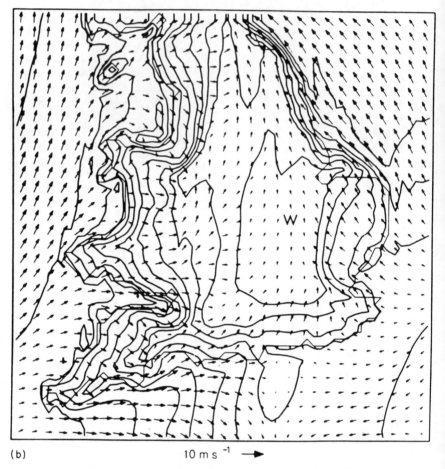

(b)   10 m s⁻¹ ——▶

Figure 16.19.   Mesoscale model predictions of surface wind and surface temperature (solid isopleths at intervals of 1 K) (a) with, and (b) without, the imposition of a sharp cloud edge. (Courtesy of Dr. K. M. Carpenter.)

the marked differences in the predictions of a mesoscale numerical model with and without the imposition of a sharp cloud edge. With the sharp cloud edge the model produces a sharp temperature gradient and sharply convergent winds changing from westerlies to easterlies over two grid squares (near the letter W). Without the cloud these changes are spread over many grid lengths and the vertical circulation is correspondingly weaker. In some cases the circulation induced by a cloud edge can lead to the triggering of significant convection. Unfortunately, numerical models do not predict the distribution of cloud cover sufficiently well to reproduce accurately its effect on the radiation balance. This is because the actual cloud cover is influenced by so many small-scale effects not represented either in the initial data or in the model formulation. Thus the pragmatic way forward is to prescribe the detailed cloud cover in the model on the basis of the satellite imagery. This can be supplemented by information on surface temperature derived from the satellite in the cloud-free areas. In order to determine the effect of the cloud on the forecast for some future time it may be desirable to define the cloud on the basis of an extrapolation of the present pattern. Clearly, however, such extrapolation forecasts are useful over only rather limited periods.

WORKSTATIONS FOR EXPLOITING RADAR AND SATELLITE IMAGERY

A theme of this chapter is the need to interpret the data from several radars and the imagery from a satellite together and in combination with other sources of information. Until recently this has been difficult because the imagery was available only in non-standard formats, often merely as hard copy. The situation is now being transformed, however, by the increasing availability of digital imagery and workstations consisting of interactive video displays which provide the following opportunities:

1. to superimpose and combine different forms of digital imagery and graphics,
2. to replay animated sequences of pictures, and
3. to enable the digital database to be modified by intervention via the video display screen.

Workstations for use in a central forecasting office

The FRONTIERS workstation for precipitation nowcasting

One of the first interactive workstations designed for operational use by forecasters is the FRONTIERS system in the UK Meteorological Office. Using concepts developed from the McIdas display system pioneered at the University of Wisconsin, FRONTIERS enables a forecaster to blend data from the several radars of the UK weather radar network along with Meteosat imagery and information from rain gauges etc. FRONTIERS is being developed for use as a central facility so that meteorologists with expertise in radar and satellite meteorology can make allowances for the unusual error characteristics of radar and satellite data and thereby produce analyses of surface precipitation and also forecasts for the period 0 to 6 h ahead. The degree of centralized quality control is intended to be such that forecasters at the outstations to which these products will be sent will be able to use them without being unduly concerned about the peculiarities of the observing techniques.

For a description of the FRONTIERS workstation the reader is referred to Chapter 3. Briefly, the system is designed to be operated on a $\frac{1}{2}$-hourly cycle, with the operator devoting nominally 10 min to the generation of each of the following:

1. a quality controlled precipitation analysis from the radar network (*product 1*);
2. a large-area precipitation analysis derived using Meteosat to extend the coverage of the radar network (*product 2*);
3. very-short-range extrapolation forecasts of the precipitation distribution (*product 3*).

In more detail, during the generation of *product 1*, the FRONTIERS operator carries out a selection of the following steps:

1. deletion of unreliable radars from the composite picture;
2. deletion of spurious echoes unrelated to precipitation;
3. monitoring and modification of bright-band (melting level) corrections derived at radar sites;
4. designation of rainfall types to allow appropriate calibration and range-dependent corrections to be applied;
5. adjustment of calibration of any suspect radars in the light of available ground truth and space/time continuity;
6. incorporation of likely orographic rainfall enhancement by specifying the nature of the low-level airflow.

The generation of *product 2* involves three main steps using the Meteosat cloud imagery. The first step is to register the imagery. The Meteosat pictures used in FRONTIERS have the distortions due to the viewpoint from space removed objectively, and they are displayed in the same national grid format as the radar pictures. However, they are not always positioned accurately, and so it is necessary to check the registration by comparison with a coastline overlay. The second step is to transform the imagery objectively to a first-guess rainfall pattern, as discussed earlier, using algorithms that relate surface rainfall to infra-red radiance and, where available, visible brightness. The FRONTIERS operator has to select either universal relationships derived for different cloud types, or current relationships based on contemporary co-located radar data. The third step is to use the rainfall pattern estimated from Meteosat to extend the rainfall analysis beyond the area covered by the radars, intervening subjectively to delete or add areas of rain inferred from the satellite so as to obtain consistency with the neighbouring radar-derived patterns and any ground truth available from other sources. There are times, especially on occasions of widespread cirrus, when the satellite guidance is positively misleading even as a first guess of the rainfall pattern.

The operator has to identify when such a situation exists, and then in such cases he has to build up his own analyses in an almost free-hand manner. The extension of coverage of estimated rainfall using Meteosat data, although clearly qualitative, is nevertheless valuable in that it sets the more accurate radar data in a broader synoptic context and gives advance warning of possible rain clouds approaching from data-sparse areas over the sea.

In the derivation of *product 3*, the rainfall forecasts, the key step is for the forecaster to determine the velocity of different parts of the rainfall pattern. One way in which this is done at present is by means of a so-called Lagrangian replay facility. With this facility the image sequence is panned so as to keep a feature of the rainfall pattern at a fixed position on the screen during the replay sequence. The forecasts are then produced automatically by applying the velocities so derived to the latest picture in the sequence.

The production of *products 1, 2* and *3* clearly has to be carried out to a very tight schedule. Therefore a design constraint placed on the FRON-TIERS display is that it should be quick and easy for a forecaster to use. It has to enable the forecaster to exercise his judgement readily within the context of a highly automated system. This is achieved by basing the system on a variety of different menus that are displayed on VDUs as required, and from which the operator can select by simply touching the screen. A typical task is one in which an area of radar echo needs to be deleted. The echo in question is defined by touching the appropriate menu item and drawing a line round the echo on the colour monitor. Similarly the displace-ment needed to position a satellite image accurately, or the adjustment of the velocity needed to optimize a Lagrangian replay sequence, can be ach-ieved by touching the menu and then using a joystick to control the position, or speed of replay, of the image(s).

An example of the use of the FRONTIERS workstation

Plates 13–18 illustrate the $\frac{1}{2}$-h FRONTIERS cycle carried out for 0400 GMT on 5 December 1983 when a cold front associated with a low over Scandinavia was moving south-east across the area of interest.

Plates 13 and 14 show the input radar composite picture and false-colour Meteosat imagery, respectively. The extensive radar echoes over central and southern England and over the English Channel are not associated with deep cloud and do not have the character of echoes from rain. A replay of a time sequence confirmed that they were not rain. By contrast, however, the echoes in the north were moving steadily south-eastwards and appeared to be rain associated with the cold front. The operator deleted the spurious echoes by using his finger to draw a line around them and selecting a delete option. In this example this was the most important part of the quality control, but some other changes, described below, were necessary to produce the final radar rainfall analysis.

The radar observations of rain have to be corrected because, being of finite depth, the rain fills the radar beam only partially at longer ranges. A correction is applied at the radar site, but this is a universal correction and does not allow for the large variations in depth from one case to another. The FRONTIERS operator applies appropriate corrections by deciding whether the rain is shallow, of medium depth, or deep. Plate 15 shows the effect that would have been achieved by selecting 'shallow', and Plate 16 shows the effect of selecting 'medium'. In choosing between these the operator is guided by his understanding of the meteorological situation and also by the red overlays which show the limits of good radar coverage (shallow rain can be observed only at relatively close range). The choice of 'shallow' would have implied that the rain at longer range was intense (coloured pink in Plate 15), and observed beyond the limits of good coverage for shallow rain. The choice 'medium' implied, more realistically, that the edge of the rain was just inside the corresponding limit of good coverage. On this occasion no other corrections were considered necessary, and the operator was able to pass rapidly through the remainder of the radar analysis.

The operator registered the Meteosat image against the coastline overlay and used a joystick to delete the shallower clouds (i.e. all except the blue areas in Plate 14) until he achieved the most credible interpretation of the cloud image in terms of surface rainfall. He then used this pattern to extend the radar data. The resulting extended rainfall analysis is shown in Plate 17. This was used as the basis of the extrapolation forecasts. The forecast for 0800 GMT (i.e. 4 h later) is shown in Plate 18. To derive this forecast the operator used the Lagrangian replay facility to determine the velocity of significant features in the imagery. He distinguished between the large mass of raincloud to the north-east, which was moving away to the east, and the remaining rainbands which were moving south-east across Britain. In reality there was some overall rotation of the rainbands, with the front travelling faster in the south, which the operator failed to take into account, but the forecast was generally quite good over England and Wales.

A workstation for initializing a mesoscale numerical model

A mesoscale model covering mainland Britain and Ireland with a grid length of 15 km is undergoing trials in the Meteorological Office. The model might be run to provide forecasts up to 18 h ahead, several times a day. Research is being carried out into the possibility of doing an hourly interactive analysis to provide a check on the last forecast issued. If the check reveals substantial deviations the forecast fields might be updated and disseminated again. One plan proposed by the Forecasting Research Branch was that each hourly interactive analysis would be followed by a 1-h model forecast to provide a first guess for the next hourly interactive analysis, etc. According to this scenario a forecaster working with the main central computer through an interactive workstation would produce analyses for the whole of the British Isles for a wide range of variables (surface pressure, temperature, humidity, wind, precipitation, cloud and visibility) using surface observations together with radar and satellite imagery.

Objective analysis schemes would be employed but the forecaster could also exploit his judgement, e.g. to incorporate discontinuities in mesoscale fields as inferred from the imagery. Only about $\frac{1}{2}$ h would be likely to be available for the task of performing each hourly interactive analysis. Therefore the interactive dialogue would have to be developed carefully, so that the forecaster could contribute as much as possible in that time. For example, the computers could be made to highlight the areas that need attention by comparing analyses with observations to show where the analysis has failed to fit the observations, or by comparing observations with first-guess fields to show up suspect reports.

A workstation for use in a regional forecasting outstation

The task of a forecaster at an outstation is to take the centrally generated products (including those generated using FRONTIERS and the mesoscale model) and use them selectively to provide forecasts for specific areas, variables and applications. This involves him in three categories of activity:

1. *Selecting* the products appropriate to his needs (thus, for example, for some application he may choose to use the FRONTIERS rainfall forecasts for 2 h ahead without modification, whilst for another application he may use the mesoscale model forecast of temperature 12 h ahead without further analysis).
2. *Modifying* aspects of the central guidance to provide an improved product for a specific requirement (thus, for example, if he requires a forecast of cloud base for 2 h ahead, he may need to carry out a combined analysis reconciling the FRONTIERS imagery, mesoscale model products and certain observations from surface stations in the light of information about the local climatology).
3. *Tailoring* the wanted information in a format suited to the needs of the customers.

A possible workstation configuration for use at an outstation is shown in Figure 16.20. It consists of (a) a data display panel, with three intelligent displays and (b) an interactive analysis display.

The data display panel in this scenario is merely a filing system with rapid

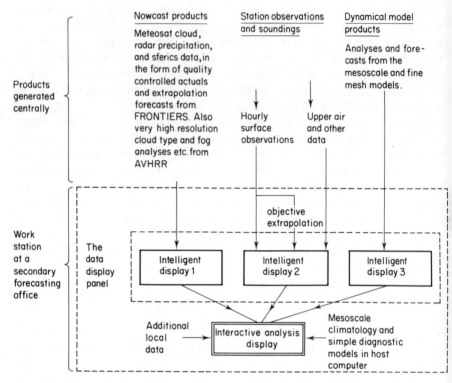

Figure 16.20.   Facilities of an outstation regional forecasting office of the future

access and easy viewing facilities. It would enable the forecaster to have simultaneous full-size displays of:

1. radar and satellite image products from FRONTIERS and other sources,
2. basic observations from reporting stations and satellite soundings, and
3. numerical products from synoptic scale and mesoscale models.

Each of the three displays in the data display panel would need to hold a sequence of charts including several actuals and hourly forecasts.

   The set of three displays constituting the data display panel would be for monitoring purposes only, and although the forecaster would be able to interact with them in the limited sense of selecting different frames and replays, he could not intervene directly to combine or modify any of the fields displayed there. To do this he would need to use the interactive analysis display in Figure 16.20. This would need to have a large number of image planes so that any required combination of the actuals and/or extrapolation forecasts could be superimposed at any brightness level, rather like stacking charts on a light table. Interaction could be by means of joystick, touchscreen and/or a data tablet, as in the case of the FRONTIERS display.

The Weather Information System (WIS) being established in the Meteorological Office will eventually be capable of delivering the necessary large volumes of digital data quickly to outstations. Workstation facilities of the kind shown in Figure 16.20 would be too expensive to be deployed at forecast outstations based on 1986 costs, but their price is coming down rapidly in real terms. Since the full potential of radar and satellite imagery, and of mesoscale numerical models, is unlikely to be realized without the kinds of facilities described here, plans are being formulated in the Meteorological Office to develop and test workstations that might be established as part of the Weather Information System of the 1990s. Similar programmes, known as PROFS and PROMIS, respectively, are under way in the USA and Sweden.

ACKNOWLEDGEMENT

I am grateful to Dr Brian Golding for helpful discussions.

Weather Radar and Flood Forecasting
Edited by V.K. Collinge and C. Kirby
© 1987 John Wiley & Sons Ltd.

CHAPTER 17

Future Development of the UK Weather Radar Network

P. RYDER AND C. G. COLLIER

INTRODUCTION

The United Kingdom weather radar network was declared operational at the beginning of 1985. This was the culmination of many years of work centred on the Meteorological Office Radar Research Laboratory (Met O RRL) at Malvern (see Chapter 1). The responsibility for the maintenance and future development of the network now lies with the Operational Instrumentation Branch (Met O 16) of the Meteorological Office based at Bracknell.

Over the past few years the experimental network of radars has grown steadily, and data have been supplied to an increasing number of meteorological office and water authority users. Building upon the software development carried out by a team from the Royal Signals and Radar Establishment, working with the Met O RRL in the early 1970s, the radar site and network software has been, and continues to be, developed by the Meteorological Office for fully operational use. This work has provided a sound basis upon which further improvements in the radar site processing and in product availability will be carried out. This chapter outlines some of the specific developments now being planned, and identifies other options currently under consideration. Users and potential users have been invited to make their needs known so that this development is as well informed as possible.

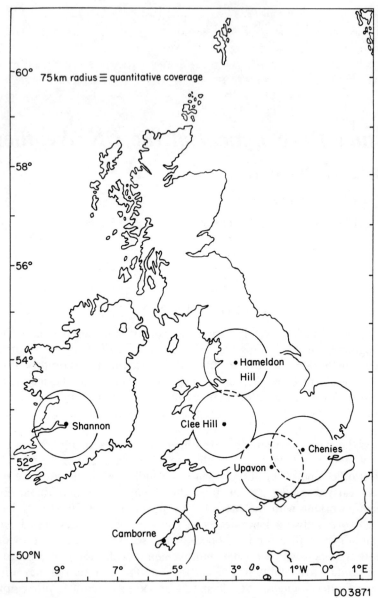

DO 3871

Figure 17.1.   Location of weather radars at May 1985

PRESENT NETWORK

Currently there are five radars in the United Kingdom covering much of
England and Wales, as shown in Figure 17.1. All the radars were manu-

factured by Plessey Electronic Systems. Camborne and Upavon are S-band systems (10 cm wavelength, 2° beamwidth), and Clee Hill, Hameldon Hill and Chenies are C-band systems (5.6 cm wavelength, 1° beamwidth). The radars at Camborne and Upavon are maintained by on-site Meteorological Office technicians because they are 15-year-old systems not designed for unattended operation. Clee Hill, although not manned, has Civil Aviation personnel close by as this installation is also over 10 years old. However, the Hameldon Hill and Chenies installations contain modern Plessey 45C radars and are totally unmanned. The reliability of these radars is now very high, as described by Hill and Robertson in Chapter 5.

Radar site data processing

Each radar site has its own dedicated minicomputer, either DEC PDP 11/40 or 11/34 systems. The software at most sites is written in CORAL-66, using DEC MACRO-11 for the time-critical areas, under the DEC operating system RSX-11M (or RSX-11S, a subset of this operating system). At Hameldon Hill the software uses a non-standard operating system and is written using the facilities of MACRO-11, although work is under way to structure the software as at the other sites. The software carries out the following main functions in real time:

1. Removal as far as possible of ground echoes and correction for partial beam blockages by obstacles, mainly hills.
2. Conversion of received power to rainfall rate.
3. Identification of the presence and height of the bright band. (The region where snow melts to form rain produces an enhanced radar echo which is referred to as the 'bright band' (see Smith, 1986).
4. Transfer from polar co-ordinates to Cartesian co-ordinates (2 km and 5 km grids).
5. Insertion of higher elevation data into the lowest elevation image to further remove ground echoes.
6. Calibration using telemetering raingauge data (see Collier *et al.*, 1983).
7. Integration over river subcatchment areas.
8. Transmission to users and the radar network centre.

Some processing is carried out within the on-site hardware, as summarized in Figure 17.2.

Network data processing

Each radar site transmits one picture every 15 min to the network centre computer, a DEC PDP 11/40, located at Bracknell. A DEC DV11 com-

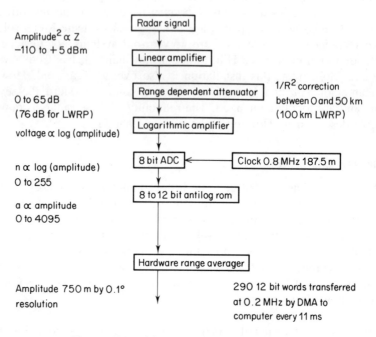

Figure 17.2.   (a) Remote site data processing

munications preprocessor receives the data from the radar sites from dedi-cated 2400 baud synchronous communications links. This equipment is cap-able of handling data from up to 16 separate channels.

Individual radar site images are stored on disc and drawn off either when data from all sites have been received or when the cut-off time for data reception is reached. A 256 × 256 × 5 km composite picture covering England, Wales and Ireland is then produced. The rainfall intensity assigned to each 5 km × 5 km square is chosen according to a hierarchy of the individual radars in operation and providing estimates for that square. The hierarchy is held in a look-up table. The chosen radar is usually, but not always, that which has the lowest beam height in the square.

The basic 256 × 256 × 5 km image produced every 15 min on the national grid is reformatted for several different outputs, namely:

1. A 256 × 256 × 5 km image output to the FRONTIERS system (Carpenter and Browning, 1984; Sargent, Chapter 3).
2. A 128 × 128 × 5 km output to the many meteorological office and water authority users distributed throughout England and Wales.

SOFTWARE

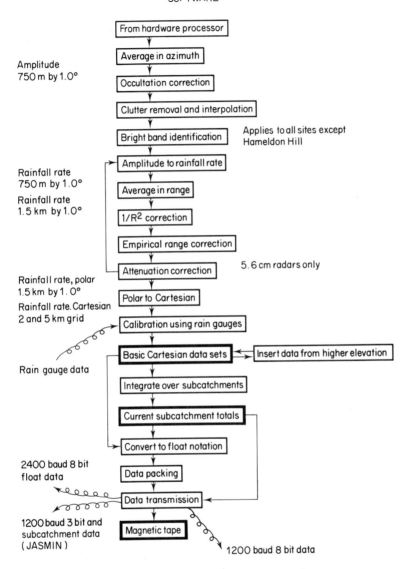

Figure 17.2 **(b)** Remote site data processing

3. An alphanumeric message for teleprinter output which is transmitted via the Telecommunications Branch (Met O 5) of the Meteorological Office, Bracknell, to those meteorological offices not possessing facilities to display colour images.
4. A 256 × 256 × 5 km image transmitted to Bracknell for insertion into

the central computer system, COSMOS, and for onward passage to the Meteorological Office OASYS (Outstation Automation System) computer systems located at London Weather Centre, Heathrow Airport, and Strike Command H/Q, High Wycombe.

As well as receiving UK radar data, the network computer also accesses data from France and the Netherlands via Bracknell every 15 min, and from Switzerland using a British Telecom autodial system, usually every 6 h but every hour if required. These data are combined with Meteosat II infra-red satellite data which are input to the network computer every hour. The combined image is reprojected in polar stereographic co-ordinates, and distributed every hour using X25 protocol to the OASYS computer system at Heathrow Airport. The image is also sent back to Switzerland every 6 h using the autodial unit, and to several other countries every 3 h as an alphanumeric message over the World Meteorological Organisation Global Telecommunications System (see Collier and Fair, 1985; Newsome, Chapter 2).

NETWORK EXPANSION

Plans are now being implemented to install a weather radar at Castor Bay in Northern Ireland, jointly financed by the Meteorological Office and several other agencies in the Department of Agriculture and Environment. The Directorate of Naval Oceanography and Meteorology is also seeking to replace old weather radars at Portland and Culdrose (the new radar is to be sited at Preddanack in Cornwall). Contracts have been placed for these three systems to be brought into operation during 1987, and for two further systems to the installed in south-west England and South Wales. A consortium comprising the Meteorological Office, Wessex Water Authority, South-West Water Authority, Welsh Water Authority and Devon County Council, which aims to install radars in the Exmoor area and South-West Wales, has been formed. Parallel negotiations are also in progress between the Meteorological Office, Anglian Water Authority, Severn–Trent Water Authority, and Yorkshire Water Authority, aimed at installing a radar in the Lincoln area. By the late 1980s weather radar coverage over England, Wales and Northern Ireland should be as shown in Figure 17.3.

Progress with installing weather radars has been slower in Scotland than in the rest of the United Kingdom: a major justification for weather radar in England and Wales—and to a lesser extent in Northern Ireland—arises from benefits to flood forecasting (NWC, 1983) and obviously analogous benefits are much smaller in Scotland. However, a study recently completed by an *ad hoc* group led by the Meteorological Office and the Scottish

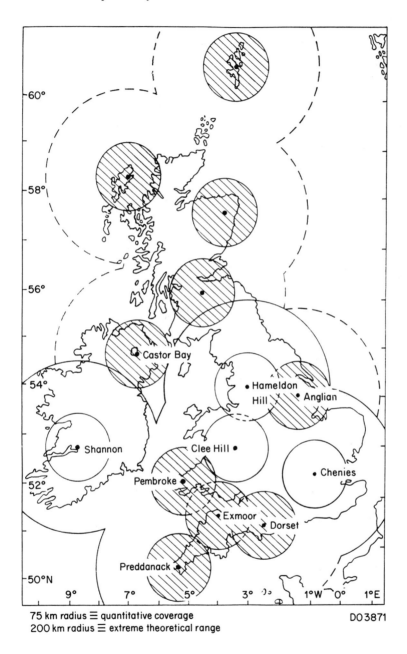

75 km radius ≡ quantitative coverage
200 km radius ≡ extreme theoretical range

DO 3871

Figure 17.3.   Location of existing and proposed weather radars

Development Department, but including a wide cross-section of the industries concerned, has identified considerable potential benefits arising from improved short-period weather forecasting, particularly for agriculture, road transport, and the building and construction industries. A network of three radars possibly located on the Outer Hebrides, in Buchan and south-west of Edinburgh has been proposed. A fourth radar on the Shetland Islands would also be beneficial to short-period forecasting for oil industry activities. These possibilities are also identified in Figure 17.3.

In order to process the data from an increasing number of radar sites, as well as to distribute new products, the radar network computer has been upgraded to a PDP 11/44 system having increased memory and disc capacity. At the same time a new network computer system has been purchased so that back-up facilities are now available for the operational services that are being provided.

DEVELOPMENT OF RADAR SITE PROCESSING

The data processing system, operating in real time at each radar site, has been developed over a number of years to provide measurements of surface rainfall which are of acceptable operational accuracy (see Collier, 1986). Nevertheless, it is believed that further improvements are possible in a number of areas, including those of ground clutter rejection and correction of the data for bright-band effects.

Ground clutter

At present the radar echoes, produced when the radar beam intersects the ground (clutter) are removed by identifying them in a stored map collected during dry conditions. No measurements of rain are attempted in those locations, and replacement values are interpolated in range from surrounding values and higher elevations. Unfortunately this can lead to the introduction of significant errors in the data even though areas of ground clutter are reduced.

When large vertical gradients of temperature and moisture occur in the lower layer of the atmosphere the radar beam may be bent downwards more than usual, causing enhanced areas of ground clutter which are not removed by the stored map technique. These anomalous conditions (often referred to as ANAPROP) produce echoes which may appear to be similar to those produced by rain. One technique which shows considerable promise for ground echo rejection (sea clutter is unlikely to be dealt with adequately) involves analysing the power returned to the radar aerial in each pulse. Rapid variations are associated with rain, and slow variations with ground clutter. Hardware has been designed and built by the Meteorological Office following principles specified by Japanese workers (see Aoyagi, 1983). It is intended to test this system on the Chenies radar before implementing it at other sites if it proves satisfactory.

Bright-band correction

Although the raingauge calibration procedure implemented at the radar sites does go some way to improving the accuracy of the radar measurements of rainfall, there remain large errors in the data when bright-band effects are evident in the radar beam close to the radar site. One solution is to employ many more calibration rain gauges, but this is uneconomic, and would only be successful with a very dense network and a considerably more complex calibration procedure.

An alternative is to specify in real time the characteristics of the bright band from the radar data themselves, and use this information to develop correction factors to be applied to the data. Such a procedure has been developed and tested experimentally with considerable success by Smith (1986). The first part of the procedure, the identification of the presence and height of the bright band, has been implemented at all radar sites except Hameldon Hill, where its implementation awaits the introduction of software written under a commercial operating system. At present this bright-band correction procedure, which identifies the intensity of the bright band and uses this with a mean vertical profile of precipitation through the bright band to derive correction factors, has not been introduced operationally. The software is complex and will require further work to absorb it into the present system without disturbing the balance in those areas where timing is critical. However, the aim is to introduce the procedure at one site, probably Chenies, in the near future.

Software structure

Although the radar site software is written mostly in a high-level language under a commercial operating system, its complexity is so great as to warrant detailed documentation. Some documentation already exists, and this has proved adequate for some time. However, the importance of introducing new algorithms and products makes it imperative to review the software structure and at the same time fully document it to commercial standards. This need is more pressing as the original expertise on which the software development was based becomes unavailable. Similarly, although CORAL-66 is a good real-time computer language, programmers with knowledge of this language are not widely available, and in the future it is likely that its commercial development will be very limited. Hence a contract is nearing completion with a software company to re-write and fully document the software in FORTRAN-77 whilst at the same time assessing the potential for expansion, so that it will be possible to identify where compromise can be made if the need arises.

Communication with radar sites

The measurement of precipitation at the radar sites involves the use of several objective algorithms. The accuracy of the resulting data is now well understood, and although improvements will undoubtedly be made it is unlikely that all error characteristics will be dealt with adequately on all occasions by purely objective methods. This fact has been recognized for some time (Browning, 1979), and has led to the development of man–computer interactive systems, such as FRONTIERS (Sargent, Chapter 3), which allow forecasters with access to a wide range of meteorological data to interact with radar data before its dissemination or use in objective forecasting procedures. The expertise available at a central location can also be used to update or modify the performance of algorithms at radar sites if two-way communications exist. At present this is not the case: the introduction of full duplex operation would considerably increase the software complexity at the radar sites and at the network centre, which the present computer memory at each site (128k words) could not sustain. However it is now possible to upgrade the memory and disc capacity significantly at the radar sites, and the time is right to consider the benefits that might accrue. Table 17.1 shows the possible areas of the radar site software which might benefit from application of meteorological information generated centrally by forecasters.

Although two-way communications with a radar site would also provide the facility for 'down-line' loading of new software, or the original software in the event of power failures (the present software already recovers from power failures automatically), these advantages are probably of secondary importance to those likely to arrive from the availability of additional meteorological information.

PRODUCT DEVELOPMENT

At present the products available from the radar network are those defined in Table 17.2. Full resolution (2 km, 5 km grids every 5 min) image data are available for water authority users as well as integrations of rainfall over river subcatchments (up to 200 per radar site) for periods of 15 min, 1 h and 24 h. Some water authority users find their present requirements satisfied by the river subcatchments defined within the radar site computers, but other users—particularly those who require integrations over very small areas

Table 17.1. Main components of the site software which might benefit from centrally generated meteorological information

Radar site software	Type of information
Raingauge calibration	(a) Weather type, including whether snow is present
	(b) Corrected radar–raingauge ratios
	(c) Additional (other gauges) calibration factors
Bright-band correction	(a) Adjustment of bright-band height
	(b) Adjustment of bright-band intensity
	(c) Correction factors for particular radial sectors
Ground/sea clutter	Extend to allow accurate removal
Range corrections (incomplete beam filling)	Appropriate corrections for particular weather types

as, for example, urban catchments—prefer to define their own integration boundaries.

The subcatchment and high resolution data are only available from the radar site. However, some users of radar data require composite images from the network centre, and, in some cases, high-resolution data from more than one radar site. At present such users must finance more than one communication link to the various sources and develop methods of using the ensuing mixture of data. The availability of FRONTIERS products in the near future, including short-period forecasts, will exacerbate the problem. The time is ripe to consider afresh how best to reconcile demanding requirements for extensive, closely tailored coverage, timeliness, high precision and spatial and temporal resolution of estimates of existing and forecast precipitation, with finite software development, processing and communication resources.

A schematic of the present system is shown in Figure 17.4. This highlights the central role of the network computer as both a data processing and communications node, and reflects the compromises which have been necessary in the generation of high-resolution single-site products and the lower-resolution composite. It is certain that any attempt to increase the number and scope of composite radar products must await the increase in processing capacity and data storage being installed at the network centre. In line with developments elsewhere within the Meteorological Office, it also seems very desirable to separate the distribution function from the product-generation

Table 17.2. Data outputs

Type 1

Consists of 5 km grid data to a range of 210 km for a colour TV display of rainfall intensity in seven levels plus zero. In addition to information on date and time and radar station number, indicators are included denoting whether the data displayed have been calibrated against rain gauge data and the type of rainfall which has been determined objectively. The data are gathered in 5-min cycles but updated for transmission every 15 min. In addition, rainfall totals for up to 200 subcatchments are provided every 15 min, with hourly totals every clock hour and daily totals every rainfall day (0900–0900 h GMT). These subcatchment totals are available for printout, display on a suitable VDU, or further processing, given appropriate hardware. At 1200 baud, the transmission time is about 35 s.

Type 2

Consists of high resolution rainfall data in 208 levels on both 2 km and 5 km grids for further processing by the recipient. The data are transmitted at 1200 baud asynchronously. The 2 km data are to a range of 75 km and the 5 km data to 210 km. These data include information on date and time, indicators to show whether the data displayed have been calibrated against rain gauge data, the calibration factors used and the type of rainfall which has been determined objectively. The objectively determined height of the 'bright band' (if one is present) is also included. In addition, this data type includes sub-catchment totals for the previous clock $\frac{1}{4}$ h, the previous clock hour and, at each clock hour only, for the previous rainfall day. Type 2 data are transmitted every 5 min, but the subcatchment data are updated only every 15 min. Transmission time is about 2 min.

Type 3

Consists of high-resolution rainfall data in 208 levels on a 5 km grid to a range of 210 km from the radar site. The data are transmitted at 2400 baud synchronously and are used by the Meteorological Office solely for compositing with data from other radars to form a picture covering a wide area for re-transmission and display. Date and time, and indicators to show whether the data displayed have been calibrated against rain gauge data, the calibration factors used and the type of rainfall objectively determined, are included. Transmission time is about 27 s.

Composite data

Consists of 5 km data to a range of 210 km suitable for a colour TV display of rainfall intensity in seven levels plus zero. Date, time and a colour code for rainfall intensity plus the height of the bright band above the radar site are included. Asynchronous transmissions at 1200 bits per second are broadcast at 5-min intervals with updates at 15-min intervals. Transmission time is slightly under $2\frac{1}{2}$ min.

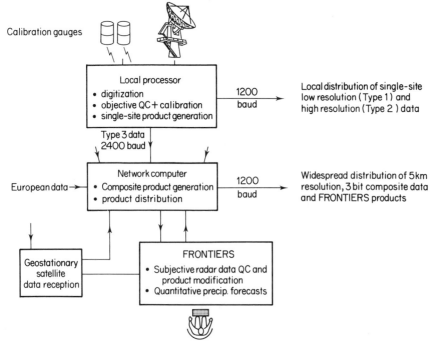

Figure 17.4. Current system for processing and distributing radar-data-based products

function at the centre. Concomitant increases in the capacity of some key communication links may not be essential, but they are undoubtedly technically feasible.

Bearing in mind the points made above, and noting the important future role of FRONTIERS and its derivatives and the likely need to increase the distribution of other information with products based on radar data, a possible schematic for a future generation and distributions system is shown in Figure 17.5. Such a system could continue to provide the single-site and composite products now available, but might also be able to generate and distribute composite products which have the higher temporal and spatial resolution inherent in the former, whilst being closely tailored to meet specific geographic requirements, using data from more than one radar when appropriate. The provision of a sequence of such estimated precipitation fields can be envisaged, which would:

(a) be based on objective quality control and available in real time;
(b) be based on subjective and further objective quality control but somewhat delayed as a result;

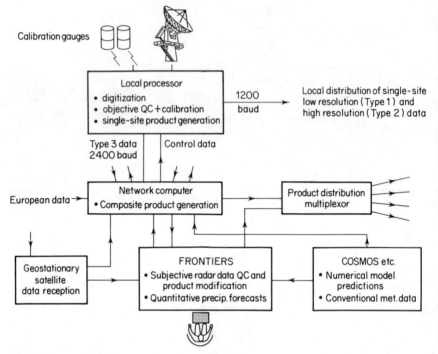

Figure 17.5.   Possible future system for processing and distributing radar-data-based products

(c) forecast precipitation fields based on the combined power of a dedicated FRONTIERS and predictive numerical models running on the Meteorological Office central computing facility, COSMOS.

Such development is compatible with foreseen improvement in the capacity of PTT analogue/modem and (increasingly) digital circuits—transmission of an $84 \times 84 \times 8$-bit field occupies a 2400 baud line for 25 s and pro rata for higher-speed lines. It should also be possible to achieve such results without major additional investment by users in their display and processing facilities. However, some further investment in central hardware and communication capacity will be necessary, and the development of the necessary system software will be a major task, requiring the updating referred to above in the first instance.

CONCLUSIONS

The weather radar network in the United Kingdom has already been developed to the stage where coverage is extensive over England and Wales.

Current plans, now being implemented, will improve the availability of quantitative data over these areas and extend coverage over Northern Ireland. A benefits study has led to the proposal for three possibly four, radars in Scotland. Clearly then the radar network will continue to expand for the remainder of this century.

Several areas of the radar site software would benefit from the real-time transmission of meteorological information from the network centre derived from the FRONTIERS system. This is practical, and could be achieved given appropriate computer enhancements.

Finally, further product development must now receive serious attention. As a first step it will be important for those concerned to agree the essential characteristics of the radar databased services and products which are likely to be required during the next few years, and a modus operandi for creating them. The very effective partnership between the Meteorological Office and some far-sighted sections of the water industry, which sustained the Dee and North-West Weather Radar projects and has played the crucial role in the creation of the UK Weather Radar network, should form the basis of that new collaboration. An early start is indicated.

REFERENCES

Aoyagi, H. (1983). A study on the MTI weather radar system for rejecting ground clutter. *Papers in Met. and Geophysics, Japan* **33**(4), 187–243.

Browning, K. A. (1979). The FRONTIERS plan: a strategy for using radar and satellite imagery for very short-range precipitation forecasting. *Met. Mag.,* **108**, 161–84.

Carpenter, K. M., and Browning, K. A. (1984). Progress with a system for nowcasting rain. Preprint Vol. *Nowcasting II*, 3–7 September, Norrokoping, Sweden, ESA SP-208, pp. 427–32.

Collier, C. G., Larke, P. R., and May, B. R. (1983). A weather radar correction procedure for real-time estimation of surface rainfall. *Quart. J. R. Met. Soc.,* **109**, 589–608.

Collier, C. G. (1986). Accuracy of rainfall estimates by radar, Part I: Calibration by telemetering raingauges. *J. Hydrol,* **83**, 207–23.

Collier, C. G., and Fair, C. A. (1985). The COST-72 pilot project: weather radar data exchange over NW Europe. Preprint Vol., Third WMO Technical Conf. on Instruments and Methods and Observations (TECIMO-III), Ottawa, Canada, 8–12 July, WMO/TD-No. 50, Instruments and Observing Methods Report No. 22, pp. 169–73.

National Water Council/Meteorological Office (1983). Report of the Working Group on National Weather Radar Coverage. National Water Council/Meteorological Office, June, 31 pp.

Smith, C. J. (1986). The reduction of errors caused by bright band in quantitative rainfall measurements made using radar. *J. Atm. and Ocean. Tech.* (In press).

Author Index

Abbott (1966), 158
Abbott *et al.* (1978), 226
Allmannspacher (1976), 193
Anderl *et al.* (1976), 192, 227, 235
Aoyagi (1983), 279
Austin and Bellon (1974), 235

Bacon (1985), 4, 7
Bailey and Dobson (1981), 157
Bastin *et al.* (1984), 235
Batten (1973), 3, 78
Boorman (1980), 147
Boussinesq (1871), 157
Box and Jenkins (1970), 140
Brandes (1975), 78
Browning (1977), 8
Browning (1979), 213, 280
Browning (1981), 72
Browning (1986), 94
Brunsdon and Sargent (1982), 131, 136
Bulman and Browning (1971), 7
Bussell *et al.* (1978), 8, 12

Cain and Smith (1976), 78
Cameron (1980), 146
Cameron and Evans (1980), 147
Carpenter and Browning (1984), 274
Central Water Planning Unit (1977), 6, 74, 85, 227, 235
Chatterton *et al.* (1979), 130
Cluckie and Smith (1980), 173
Collier (1984), 3, 4, 236
Collier (1985), 174
Collier (1986a), 88, 93, 278
Collier (1986b), 90, 93
Collier and Cluckie (1985), 184
Collier and Fair (1985), 21, 276
Collier and Knowles (1986), 94
Collier and Larke (1978), 174
Collier *et al.* (1975), 6

Collier *et al.* (1980a), 71
Collier *et al.* (1980b), 81
Collier *et al.* (1983), 71, 78, 81, 83, 88, 224, 273
Creutin and Obled (1980), 141

Damant *et al.* (1983), 215

Ede and Cluckie (1985), 177
Evans (1980), 131
Evans (1981), 214
Eyre and Crees (1984), 136

Grimshaw and Wang (1980), 136

Hamlin (1983), 226
Harpin (1982), 173, 177
Harrold *et al.* (1974), 6, 72, 85, 235
Hill (1983), 84
Hill *et al.* (1977), 235
Hill *et al.* (1981), 84
Hydraulics Research Station (1981), 146

ICA Steering Group (1985), 218
Institution of Water Engineers and Scientists (1982), 218

Jazwinski (1970), 224
Jones (1980), 155
Jones and Moore (1980), 234
James (1981), 24
Joss *et al.* (1970), 74

Klatt and Schultz (1985), 192
Krietzberg (1981), 213

Lambert (1972), 134, 145
Lambert and Lowing (1980), 136
Lambert and Reed (1986), 136

Lighthill and Witham (1955), 158
Linsley *et al.* (1949), 158

Matsubayashi *et al.* (1984), 230
McCarthy (1983), 158
Manley *et al.* (1980), 155
Moore (1977), 225
Moore (1980), 145
Moore (1982), 140, 234
Moore (1983), 234
Moore (1985), 230
Moore and O'Connell (1978), 145
Morris (1980), 226
Morris (1983), 140

National Water Council (1981), 215
National Water Council —
    Meterological Office (1983), 11,
    213, 276
NERC (1975), 134, 147, 172
Negri and Adler (1981), 26
Newsome (1981), 21
Ngirane-Katashaya and Wheater
    (1985), 226
Niemczynowicz (1984), 226
North-West Water (1977–1980), 144
North-West Water (1979–1980), 145
North-West Water (1980), 147
North-West Water (1984), 145
North-West Water Authority (1985), 10
NWWA (1983), 215
NWWA (1984), 218

O'Donnell and Groves (1979), 144
Owens (1986), 173, 179

Plackett (1950), 177
Poisson (1816), 157
Price (1977), 158, 161

Reed (1984), 1365, 234
Rodda and Flanders (1985), 213

St Venant (1871), 158
Samuels and Gray (1982), 158
Schultz (1969), 192
Schultz and Plate (1976), 206
Severn-Trent Water Authority (1986),
    16
Sherman (1932), 172
Singh (1977), 224
Singh and Woolhiser (1976), 224
Smith (1986), 90, 91, 273, 279

Taylor (1975), 8
Taylor and Browning (1974), 7, 8
Troutman (1982), 225
Troutman (1983), 225

Walker *et al.* (1983), 214
Walsh (1984), 218
Water Research Centre (1983), 215
Water Resources Board (1973), 8
Wilson (1970), 78
Wilson and Brandes (1979), 78, 85

Yoshino (1985) 235

Subject Index

Accuracy, 4, 5, 7, 28, 50, 71–95, 130, 148, 228, 236, 278
  seasonal variations, 88
Adaptive forecasting, 139, 194, 195
Advection, 35, 43
Air Ministry, 4
Anenometer data, 84
Anglian Water Authority, 276
Anomalous propagation (ANAPROP), 48, 50, 60, 72, 83, 278
Areal rainfall, 5, 16, 19, 109, 114, 120, 192, 215, 223, 224, 226, 233, 235
ARMA Model, 145, 148
Assessment factor, 72, 74–76, 78, 82, 84, 88, 93
  temporal variability, 81
Attentuation, 7, 36, 39, 60, 72
Automated data capture, 211, 216, 218, 219
Automatic weather stations, 99

Background field, 260
Bala lake, 5
Bankfull discharge, 162, 163, 165, 166
Baroclinic leaf cloud, 257
Baroclinic zone, 257
Base flow, 145, 149, 161, 172, 194
Base station receiver, 124
Basin-wide flood forecasting, 154
Bavarian Institute of Water Resources, 206
Beam width, 5
BEST EST, 48–50
Birmingham University, 145, 151
Bispectral information, 242
Boussinesq equation, 227
Bright-bank, 6, 41, 45, 48, 60, 71, 78, 80, 81, 83–85, 88, 89, 93, 94, 273, 278, 279, 281

British Telecom autodial system, 276
Calibration of weather radar, 9, 47, 71, 72, 74, 78, 81, 83–85, 93, 94
  on-site calibration, 47
Cardiff Weather Centre, 15
Catchment data, 16
Celyn reservoir, 5
Central Forecasting Office, 35, 45
Central Water Planning Unit, 6, 8, 10, 55, 189
Channel storage, 162, 163
Chester Conference, 8, 9, 16
Civil Aviation Authority, 9
Climatological (cloud/rain) transfer functions, 242, 243
Cloud boundaries, 261
Cloud cover and temperature, 19, 42
Cloud imagery, 239, 241, 242, 247, 254, 263
Cloud physics, 4
Clutter cancellation, 36, 39, 40, 41, 56, 60
Coastline overlay, 264, 266
Cold pool, 247
Communications system, 110, 111, 112, 113, 123, 126, 273, 274, 280, 281, 284
  x25, protocol, 276
Composite radar, 7, 12, 16, 21, 29, 36, 45, 48, 62, 118, 186, 264, 265, 274, 281, 282, 283
Computer systems, 12, 24, 111
  BBC microcomputer, 17
  COSMOS, 276, 284
  DEC DV11 communications preprocessor, 274
  DEC PDP11/34, 62, 69, 111, 273
  DEC PDP11/40, 25, 273
  DEC PDP11/44, 278

DEC VAX 11/750, 38
Delta Technical Services TG4000
    processor, 63
Digivision colour monitor, 64
IBM Series I minicomputer, 166
ICL 2900 series, 113
Jasmin store, 45, 64
OASYS (Outstation Automation
    System), 276, 284
Plessey minicomputer, 166
RAMTEX graphics device, 38
Conceptual hydrological models, 129,
    134, 153, 154, 157, 234
Conceptual life-cycle models, 253, 256
Conceptual models of weather systems,
    241, 243, 244, 246, 256, 259
Conditional probability rainfall forecast,
    197–202
Contributing area of storm runoff, 225
Control of water resource systems, 214
Convective rainfall, 26, 41
Convergence of winds, 102
Committee for tidal gauges, 123
Cost 72 project, 1, 19–33, 206, 239
    Assessment of utility, 27
    Production of image, 24–27
Cyclogenesis, 257

Darmstadt, 124, 125
Data archiving, 47–51, 62, 124, 131,
    211, 212, 214, 216, 217, 218
    collection (capture) systems, 211,
        212, 213, 214, 215, 218, 220
    communication systems, 9, 24, 109,
        157, 211, 212, 217, 218, 219, 220
    collection platform (DCP), 124
    display panel, 267
    dissemination, 213, 216
    loggers, 123
        sold data, 113
        interrogable, 124, 168
    management, 218
    network, 273
    presentation, 220
    processing, 8, 9, 12, 15, 47–51, 62,
        69, 90, 167, 214, 215, 217, 281,
        283, 284
    retrieval, 217, 220
Database, 212, 220, 236, 263
Data-ring, 218

Decca Radar Limited, 3, 4
Decision making, 202, 206, 213, 214,
    215
Dee and Clwyd River Authority, 5, 6
Dee Weather Radar and Real-time
    Hydrological forecasting project,
    4–9, 15, 55, 74, 89, 227, 235, 285
Deep-convective phenomena, 245
Defford, 8
Degraded data, 16, 171, 183, 184, 189
Degree-day method, 140
Delta factor, 173, 179, 183, 184
Department of Agriculture and
    Environment, Northern Ireland,
    276
Design flood estimation, 132, 134
Devon County Council, 276
Diabatic heat source, 259
Digital
    circuits, 284
    imagery, 263
    terrain models, 141
Directorate of Naval Oceanography
    and Meteorology, 276
Displacement, 44
Dissemination of radar/satellite data,
    21, 22
Distributed models, 227, 228
DODO conceptual routing model, 158,
    161
Domain calibration procedure, 84–88,
    93, 177
Doppler radar, 3, 5
Drop size of rain, 41
Dynamic flood plains storage, 165

East Hill, Dunstable, 4
Empirical state adjustment, 224, 234
Error prediction, 129, 136, 139, 140,
    145, 166, 233
European centre for medium range
    weather forecasting, 20
Echo
    permanent, 6
Echo signal analyser, 5
Effective rainfall, 144, 172
Encoders, 113
Extrapolation forecasting of weather
    systems, 248, 249, 250, 252, 253,
    256, 259, 263, 264, 266, 269

Farnborough, 4
Federal Republic of Germany, 191, 206
Fine-mesh model, 257, 259
Flapped outfall, 163, 165
Flood
 alleviation, 130, 164, 167
 damage, 11
 duty rota, 122
 economic viability, 125
 forecast room, 117
 peak flow time, 121
 risk zones, 115, 117, 120, 121, 123
 'stand-by' and 'alarm' messages, 115
 warning, 11, 14, 15, 17, 31, 109–126,
  129, 143, 153, 191, 195, 202,
  206, 212, 225, 232, 235
 schedules of action, 115–117, 129,
  130, 155
FLOUT, river catchment model, 146,
 148
Forecasting, 4, 6, 8, 11, 17, 19, 43, 44,
 45, 55, 90, 97, 101, 105, 122, 132,
 141, 143, 155, 168, 172, 186, 191,
 192, 195, 202, 203, 213, 234, 235,
 240, 263, 267, 278, 281, 284
 cloud base, 267
 dissemination, 121
 flow, 6, 11, 15, 16, 17, 31, 55, 56,
  94, 102, 118, 120, 129–152,
  153–169, 171, 191, 213, 236
 precipitation, 264, 266
 temperature, 107, 267
 tide surges, 117, 118, 121
Forth River Purification Board, 136
France, 276
Franklaw, 58, 62, 64, 111–113, 114,
 117, 118, 143
Fronts
 ANA type, 106
 KATA type, 106
Frontal cloud system, 250
 rainfall, 26
FRONTIERS, 11, 12, 35–46, 94, 144,
 145, 169, 172, 184–189, 196, 211,
 213, 218, 263, 267, 274, 280, 284,
 285
 accuracy, 45
 display, 265
 future developments, 45
 offline use, 45

 operational use, 38–44
 operator, 266
 products, 264, 265, 281
 system hardware, 38
 workstation, 263–266, 268–269

Geostationary satellite, 239
German Research Foundation, 206
German Weather Service, 193, 202, 206
GOES imagery, 255
Greater London Council, 12
Grid square model, 227, 235
Ground echoes (clutter), 72, 83, 194,
 273, 278, 281
Ground truth, 61, 264
Groundwater, 227, 229, 234

Haddington flood warning system, 136
Harmonic analysis, 81–83
HOMS, 136
Horizontal divergence field, 259, 260
Humidity analyses, 260
Hydraulics Research Station, 146
HYREUN model, 192, 194

Impulse response, 172, 173, 177, 180,
 181, 184
Infiltration
 capacity, 228, 229
 parameter, 194
Information
 needs, 215
 systems, 216, 217
 technology, 211, 214, 215, 219
Infra-red imagery, 37, 42, 43, 242, 248,
 254, 264, 276
Institute of Hydrology, 144
Institute of Oceanographic Sciences,
 124
Instrumental variables algorithm, 177
Intensity distribution of rainfall, 6, 71
Interactive video display, 263
Isentropic surface, 258
ISO function model, 134–136, 139, 145,
 148, 149
Isolated event model, 134–136, 137,
 138

Japan, 230, 279
Jasmin processor, 15, 16, 17, 45, 97, 99

Joystick, 44, 265, 266, 269

Kalman filter, 146, 224, 234
Katabatic winds, 102

Lag and route methods, 158
Lag time of floodwave, 161, 162, 163
Lagrangian replay facility, 44, 265, 266
Lancashire Conjunctive Use Water
  Supply Scheme, 109, 111, 214
Lancaster University, 144, 239
Latent heat effects, 260, 261
Lateral inflows, 153, 156, 159, 160,
  232, 233
Level slicing of radar/satellite data, 26,
  184
Line convection, 243, 244, 252, 254,
  260
Logica Ltd, 35
Logica Vitesse System, 16

Magnetic AB (Sweden), 62
Malvern, 4, 6, 7, 8, 16
Management information system, 215
Manchester Weather Centre, 97, 99
Map projection of radar data
  polar-stereographic map projection,
    192, 273, 276
  UK National Grid Projection
    (Transverse Mercator), 24, 242,
    264, 273, 274
Marconi, 4
Master station, 218
McIda display system, 263
Mersey Basin forecasting system, 144,
  147
Mesoscale
  convective system, 242, 245, 246,
    253, 254, 255, 256
  numerical models, 259, 260, 261,
    262, 266, 267, 268, 269
  weather phenomena, 240, 241, 243,
    247, 251, 253
  workstation for initialization, 266
Meteorological Office, 4, 5, 7, 8, 10,
  11, 12, 13, 55, 88, 90, 94, 99, 114,
  118, 122, 125, 143, 172, 184, 209,
  257, 260, 263, 266, 269, 270, 273,
  285
  Advisory Services Branch (Met 03)
    47
  Forecasting Research Branch, 267

Operational Instrumentation Branch
  (Met 016), 48, 270
Radar Research Laboratory (Met O
  RRL), 271
Telecommunications Branch (Met
  05), 275
Meteosat, 239, 240, 242, 254, 255, 256,
  257, 263, 264, 265, 266
  data, 8, 24, 25, 37, 38, 42, 117, 276
Microconsultants Limited, 62
Microwave radio link, 64
Mid-level convective areas, 252
MINILOG, 50
Ministry of Agriculture, Fisheries and
  Food, 10, 11, 55, 141, 189, 236
Missing data, 144, 147, 157, 166, 168,
  174, 225
Models for flood forecasting, 17, 31,
  109, 111, 114, 118, 129, 144, 153,
  171, 223
  accuracy, 114, 191, 206, 223
  benefit, 114, 130, 140, 191, 202, 223,
    224
  bias, 225
  calibration, 134, 145, 153, 166, 167,
    171, 177, 195, 225, 230, 235, 236
  correlation/graphical methods, 133,
    143, 145, 146, 148, 150
  distributed models, 192, 226, 228
  evaluation, 236
  event-based, 171, 172
  future developments, 209
  geometrically distributed (models)
    223, 228
  grid-square models, 226, 227, 235
  hydraulic methods, 133
  lead-time correction, 114, 118, 148
  linear models, 224
  model order, 179
  non-linear models, 224
  operational use, 147, 148
  probability-distributed models, 223
  probabilistic forecast, 141
  robustness, 223
  sampling intervals, 179
  ungauged catchments, 132, 147, 233
Muskinghum-Cunge routing, 129, 133,
  146, 148, 158, 161
  variable parameter, 133, 161

National Water Council, 11
Netherlands, 276

Net rainfall, 135
Networks
  hydrometric, 220
  radar, European, 1, 8, 19–33, 206, 239
    future development, 271–285
    UK, 1, 7, 8, 11, 17, 209, 239, 263,
      271, 284, 285
  raingauges, 6, 9, 19, 226, 233, 235
  river gauges, 216
North Channel
  gap, 250
  shower bank, 244, 249, 253
North West Radar Project, 7, 9, 11,
  53, 55, 71, 72, 171, 174, 189, 211,
  214, 215, 217, 218, 220, 285
North West Water Authority, 9, 10,
  14, 55, 56, 88, 90, 92, 93, 94, 99,
  105, 109, 112, 118, 121, 125, 136,
  143, 144, 167, 171, 174, 175, 179,
  189, 211, 212, 213, 215, 217, 219,
  220
  Lancashire and South Cumbrian
    Division, 105
  Mersey and Weaver Division, 102
  regional communications system, 11,
    55, 56, 148
North West Weather Radar
  Consortium, 145, 189
Northern Ireland, 285
Nowcasting, 240, 241
  development facility, 270
  extended nowcasts, 241
  precipitation nowcasting, 263
Numerical-dynamical models, 240, 284
  adjustment in light of imagery, 256,
    257
  initializing, 241, 259
Numerical weather prediction, 240

Occultation, 36, 39, 40, 48
On-line forecasting, 6
Operating system
  DEC RSX-11M, 273
  DEC RSX-11S, 273
Operational assessment centres (OAC),
  21, 27
Operational experience, 53
Operations Systems Group, 8
Orographic rainfall enhancement, 41,
  42, 43, 44, 72, 74, 78, 81, 99, 264
Out-of-bank flood, 161, 162, 163, 165
Outstations, 110, 113, 123, 124, 218,

267, 269
  raingauges, 113
  river level, 113
Overland flow, 229

PAL encoder, 15
PARAGON, 47, 51
Parameter adjustment (updating), 129,
  136, 140, 145, 173, 192, 194, 195
Pennine hills, 72, 84
Physics-based models, 140, 172, 226
Plessey Electronic Systems, 273
Plessey (Radar Limited) Company, 4,
  5, 7, 17
Potential vorticity, 257, 258
PPT analogue/modem, 284
Pre-event flow, 136
Precipitation, analysis, 264, 266
Primary Data User System (PDUS), 37
Probability density function, 228, 229
  exponential, 229, 230
  gamma, 232
  Gaussian, 232
Probability-distributed rainfall/runoff
  models, 223, 228, 230, 232
Punched-tape recorders, 113
Pure time delay, 134, 136

Radar
  accuracy, 4, 5, 35, 71–95, 174, 194,
    241, 242, 280
  antenna, 62
  beam elevation, 57, 58, 85
  benefits, 8, 11, 12, 14, 21, 31, 56,
    90, 107, 213, 235, 276
  calibration, 16, 36, 39, 41, 47, 71,
    72, 193, 194, 242, 279, 281
  communications, 36, 62, 273
  correction, 39, 40, 60
  cost, 4, 5, 6, 7, 11, 12, 21, 55, 65, 90
  coverage, 12, 13, 14, 36
  data,
    high resolution, 281
    subcatchment, 280, 281
    Type 1, 2, and 3, 282
  Data Users' Quarterly Liaison
    Report, 50
  development, 3, 241
  echoes, 4, 40, 41, 56, 60, 83, 101,
    264, 265, 273
  evaluation, 56, 235
  future technology, 209

fault monitor, 64, 65, 69
grids, 12, 24, 61, 192, 243, 261
improvements, 69
preventive maintenance, 64, 65
resolution, 37, 47, 184, 280
signal averaging unit, 62, 69
site selection, 56, 57
software, 12, 15, 19, 69, 271, 273,
    284
  failures, 65
swept gain unit, 69
Radar Research Station, 4
Radar sets
  C band, 4, 5, 7, 9, 273
  Plessey type 45C, 62, 66, 223, 273
  S band, 4, 5, 7, 9, 273
  Type 40X band Storm Warning, 4
  Type 43S, 4, 5, 7
  Types 40, 41, 42 and 43X, 4
  Types 45C, 45S, 46C, 7
  X band, 4, 7
  sites
    Cambourne, 8, 9, 16, 17, 40, 273
    Castlemain, 8
    Caster Bay, Northern Ireland, 276,
      277
    Chenies, 12, 15, 41, 69, 273, 279
    Clee Hill, 9, 15, 16, 41, 91, 273
    Culdrose, 276
    Dorset, 277
    Exmoor, 15, 17, 276
    Hameldon Hill, 9, 11, 15, 16, 41,
      57, 62, 67–69, 81, 85, 90, 93,
      99, 113, 118, 143, 174, 176,
      185, 186, 239, 273, 279
    Lincoln, (East Anglia), 16, 276
    Llandegla (Dee), 5, 8, 16
    Pembroke, 15, 17
    Portland, 276
    Predderack, Cornwall, 276, 277
    Shannon Airport, 12, 40
    Upavon, 8, 9, 16, 40, 273
Radiation balance, 261, 262
Radio-sonde, 98, 99, 100
Rain cells
  development and heavy, 240
RAINBOW, 4
Raindrop
  growth, 71
  size, 71
Rainfall
  average, 1

errors, 224, 228
excess (effective), 224, 235
in semi-arid regions, 236
intensity classes, 61
  time distribution, 202–204
measurement by radar, 57, 60
pattern, 226
space-sampling errors, 225
spatial
  interpolation, 235
  histogram, 230, 231
  sampling of, 226
  variation, 223, 226, 228
Rainfall runoff models, 109, 114, 118,
    121, 126, 129, 132, 133, 134, 136,
    140, 143–145, 148, 153, 156, 157,
    160, 171, 172, 191, 192, 194, 202,
    223, 224, 225, 228, 232, 233, 234,
    235
Rainfall separation techniques, 134, 194
  type, 41, 74, 81, 83, 93, 264
  warning, 99, 100, 102, 105, 118
Raingauge sites
  Abbeystead, 74, 75, 76
  Alwen, 74, 75
  High Beatham, 89
  Hollingworth Lake, 81
  Swineshaw, 74, 77, 78, 81
Raingauges for calibration, 72, 85–87,
    113
Real-time correction (updating
    techniques), 136, 166, 172, 173,
    194, 195, 223, 225, 233
  system, 5, 129, 131, 132, 153
Recursive-least squares algorithm, 177,
    179
Recursive parameter estimation, 173
Reflectivity gradients, 81, 90
Regional communications scheme
    (RCS), 109, 125, 126, 143, 148
Regional forecasting outstation, 267
  workstation for, 267
Regional forecasting service, 97–107
Reservoirs
  control rules, 6, 214
  operation, 121, 191, 202, 206, 235
  regulation, 5, 16, 56, 213
Resource allocation, 214
Richards equation, 227
River
  basin management, 212
  Danube, 193

Darwen, 149, 151
Dee, 5, 8, 74, 136, 227
Gung, 193, 202, 203
Hirnant, 241
Irwell, 171, 176
Ribble, 148, 149
Severn, 163, 167
Vyrnwy, 163
Wyre, 149, 150
River level flood warning, 133
Routing models, 109, 114, 118, 121,
    129, 132, 133, 134, 136, 140, 145,
    146, 153, 156–168, 215, 224, 227,
    232, 233
Royal Signals and Radar
    Establishment, 4, 7, 8, 35, 271
Ruhr University, 207
Runoff
    production, 228, 229
    proportion, 134, 135, 145, 157, 172,
        173, 174

St Venant equations, 134, 157, 158
Sampling rate, 72, 171, 193
Satellite
    communications, 124
    data, 1, 8, 12, 19, 21, 25, 37–38, 42,
        43, 99, 117, 206, 209, 240, 243,
        255
    calibration, 243
    cloud/rain transfer functions, 252
    imagery, 241, 244, 252, 256, 257,
        262, 267
    forecast data, 43, 44
    radar/satellite correlation, 42, 43,
        45
    satellite slicing technique, 43
Satellite Meteorology Branch, 35
Science and Engineering Research
    Council, 189
Scotland, 285
Scottish Development Department,
    276, 278
Semi-distributed catchment model, 136,
    140
Sensors, 217, 218
Severn and Trent river basins, 140,
    153, 154, 158, 166
Severn–Trent flow forecasting system,
    166
Severn–Trent Water Authority, 15, 132
    153, 155, 276

Sewers
    stormwater overflows, 215
    subject to flooding, 123
Sewerage
    pump control, 214
    systems, 215, 220
Shallow layer cloud, 41
Sharp cold front, 243
Sharpening gradients, 260
Short-period Weather Forecasting Pilot
    Project, 8
Single-site radar data, 48
Smoothing factor, 173
Snow, 41, 56, 71, 174, 194
Snowdonia, North Wales, 84
Snowmelt forecasting, 133, 140
Software
    CORAL-66, 273, 279
    DEC MACRO-11, 273
    documentation, 279
    FORTRAN-77, 280
Solenoidal circulation, 261
Southern Water Authority, 12, 15
South West Water Authority, 17, 276
Squall line, 106, 107
State adjustment (updating), 129, 136,
    140, 234
Static floodplain storage, 163, 165, 166
Steady-state gain, 172, 173, 179, 189
Stochastic rainfall forecasting, 191, 192,
    195, 196
Storage capacity, 228, 229, 232, 234
Storage models, 129, 134–136, 227
Storm
    cell, 249
    cluster, 249
    movement, 226
    tide warning service, 117, 118, 123
Subgridscale convection, 260
Subsynoptic
    comma cloud system, 254, 255, 259
    positive vorticity advection centre, 243
Surface runoff, 234
Switzerland, 276
Synoptic
    chart, 97, 105
    forcing, 259
    reporting stations, 97, 98, 99
    weather patterns, 99, 100
Synoptic-scale
    dynamical systems, 240
    frontal system, 251

Telemetry system, 15, 17, 111, 131,
    156, 166, 211, 214, 217, 218, 219,
    220
Thames Water Authority, 1, 2, 15
Thunderstorm
  development, 255
  dynamics, 5
Tidal
  flooding, 117, 118, 121
  stations, 111
Time-area diagram, 136, 192
Time of concentration, 160
Time of travel, 227
Tipping bucket raingauge, 113
Toppographically forced phenomena,
    244, 253
Touchscreen, 268
Transfer function model, 129, 134, 144,
    145, 148, 171, 172, 177, 179, 189,
    192, 225, 227
Travel time of flood peaks, 133
TREND message, 106
Tropopause, 245
Troposphere, 248, 251, 258
Turbulence

USA, 218, 255
Unit hydrograph models, 129, 134, 143,
    144, 147, 172, 177
Unmanned weather radar, 7, 9, 55, 65
Urban flooding, 155, 226
User requirements of radar data, 21, 23
Users of radar data, 6, 14, 32, 56

Velocity, 3, 44, 240, 265, 266
  vertical distribution of, 259
Visible imagery, 242, 264

Wallingford procedure, 215
Warm sector of depressions, 78
Warrington, 111, 114, 117, 118
Washband storage, 163
WASSP model, 215
Water Act (1973), 6
Water Data Unit, 6
Water management, 5, 213
Water Research Centre, 6, 8, 10, 55,
    144, 146, 189, 218
Water Resources Board, 5, 6
Water services, 215, 216
Water vapour imagery, 242, 247, 248,
    257
Wave speed, 161, 162, 166
Weather Information System (WIS),
    269, 270
  PROFS, 270
  PROMIS-90, 270
Welsh National Water Development
    Authority, 6
'Welsh shadow effect', 78, 83, 84
Welsh Water Authority, 14, 15, 136,
    276
Wessex Water Authority, 16, 136, 276
Wind speed and direction, 74, 76–79
Wisconsin University, 263
Workstations, 263–270
World Meteorological Organization,
    136
  Global Telecommunications System,
    276

Yorkshire Water Authority, 15, 57, 62,
    64, 114, 276

Z.R. relation (reflectivity/rain rate), 41,
    57, 71, 72, 74, 78, 84, 85, 194